MW00814564

Springer Series in Optical Sciences

Volume 180

Founded by
H. K. V. Lotsch

Editor-in-Chief
W. T. Rhodes

For further volumes:
http://www.springer.com/series/624

Springer Series in Optical Sciences

The Springer Series in Optical Sciences, under the leadership of Editor-in-Chief William T. Rhodes, Georgia Institute of Technology, USA, provides an expanding selection of research monographs in all major areas of optics: lasers and quantum optics, ultrafast phenomena, optical spectroscopy techniques, optoelectronics, quantum information, information optics, applied laser technology, industrial applications, and other topics of contemporary interest.

With this broad coverage of topics, the series is of use to all research scientists and engineers who need up-to-date reference books.

The editors encourage prospective authors to correspond with them in advance of submitting a manuscript. Submission of manuscripts should be made to the Editor-in-Chief or one of the Editors. See also www.springer.com/series/624

Ken-ichi Shudo · Ikufumui Katayama
Shin-ya Ohno
Editors

Frontiers in Optical Methods

Nano-Characterization and Coherent Control

 Springer

Editors
Ken-ichi Shudo
Yokohama National University
Yokohama
Japan

Ikufumui Katayama
Shin-ya Ohno
Faculty of Engineering
Yokohama National University
Yokohama
Japan

ISSN 0342-4111   ISSN 1556-1534   (electronic)
ISBN 978-3-642-40593-8   ISBN 978-3-642-40594-5   (eBook)
DOI 10.1007/978-3-642-40594-5
Springer Heidelberg New York Dordrecht London

Library of Congress Control Number: 2013953268

Printed on acid-free paper

Springer is part of Springer Science+Business Media (www.springer.com)

Preface

The field of spectroscopy emerged in the seventeenth century when Isaac Newton discovered that sunlight consists of many different colors. The next milestone came when Maxwell formulated the theory of electromagnetism, in which he attributed differences in color to differences in wavelength. Furthermore, he predicted that electromagnetic waves can have any frequency (or, equivalently, wavelength) and that visible light itself is a kind of short wavelength electromagnetic wave. Since this discovery, the frequency range accessible by spectroscopic methods has expanded from a few Hz up to PHz (10^{15} Hz), and beyond. Recent developments in electrical and optical technologies have enabled new forms of spectroscopy to emerge, such as nonlinear spectroscopy. Spectroscopic techniques are no longer restricted to homogeneous gaseous, liquid, or solid materials, but have expanded to accommodate nanometer-scale materials including nanoparticles, thin films, and nano-composites.

This book, "Frontiers in Optical Methods: Nano-Characterization and Coherent Control," covers a broad range of recent advances in spectroscopy, especially in nano-materials and under extreme conditions. Spectroscopy basically involves the generation of light (electromagnetic waves) and the detection of material responses. Recent developments in the former area have been facilitated by technological advances involving two key light sources—lasers (especially ultra-short pulse lasers) and relativistic radiation from accelerators. Both sources provide intense and controllable broadband radiation, from the terahertz to X-ray ranges, and have become ubiquitous tools in the investigation of materials' properties. In the latter area, recent developments in electronics and imaging devices, as well as methods for improving sensitivity under extreme conditions (such as at high pressures or low temperatures) have provided important new insights into properties of materials that cannot be studied using conventional light. Within this book, you will find out how these techniques are applied in modern spectroscopy.

The main parts of this book are devoted to three major topics: I. Reflectance Spectroscopy, II. Ultrafast and Coherent Measurements, and III. Terahertz Technology.

In Part I, we discuss the extreme sensitivity of spectroscopy at visible and ultraviolet wavelengths, which is capable of detecting sub-monolayer films on well-characterized surfaces (K. Shudo and S. Ohno). The paper provided by

T. Nanba reviews microscopic spectroscopy at extremely high pressures. S. Kimura discusses the powerful accelerator light sources, and their role in broadband material spectroscopy.

In Part II, we summarize the recent advances in ultrashort-pulsed laser spectroscopy. Since the pulse duration can be made shorter than the timescale of various elementary excitations in materials, we are concerned with coherent dynamics. The impulsive excitation of solid materials by such lasers can induce coherent lattice vibrations (see the papers from K. Kato and O. Matsuda) and pulsed accelerator light sources enables the full characterization of the laser-induced atomic displacements (Y. Tanaka).

In Part III, coherent terahertz spectroscopy, which has rapidly increased in popularity, is reviewed (M. Tonouchi). Material characterization using ultrashort-pulsed laser-based terahertz sources is discussed by I. Katayama and K. Kitagishi. In combination with accelerator sources, the light intensity can be significantly enhanced (review by M. Kato), which is useful for nonlinear and microscopic spectroscopy. Recent electronic developments have also provided important breakthroughs in detection sensitivity, even to the point of a single terahertz photon (K. Ikushima).

This book thus presents a range of modern applications of advanced light sources as material probes, notably on the nanometer scale. Although the research presented in this book deals with specific topics, the underlying concepts are essential for the spectroscopic analysis of nanoscale materials and for spectroscopy at ultrafast timescales. The combination of these light sources with the sophisticated techniques presented can confidently be expected to expand the fields of nanoscale-, bio-, and complex materials spectroscopy, and to greatly enhance our understanding of quantum phenomena, especially in extreme conditions. It is our great hope that this book will thus help to expand the frontiers of materials science and its applications to new directions for the future.

Finally, this book is inspired by a review volume of the Journal of the Vacuum Society of Japan (in Japanese), and a number of figures are reprinted from the volume with permission by courtesy of the Vacuum Society of Japan. We would also like to thank Claus E. Ascheron and his coworkers at Springer-Verlag in Heidelberg for their help in completing this book.

Yokohama, Japan   Ken-ichi Shudo
                                                              Ikufumui Katayama
                                                                  Shin-ya Ohno

Contents

Chapter 1
Introduction: Ultra-Fast Response of Ultra-Thin Materials on Solid Surfaces

Ken-ichi Shudo

1.1 Introduction

1.1.1 Photoabsorption

"Something that makes things visible"
This is a dictionary definition of *light*. Originally a broad and ill-defined idea, in the modern parlance *light* usually refers to light that is visible (Vis) to the human eye, with a wavelength in the range of approximately 400–750 nm. Since light is understood as the propagation of *any* electromagnetic radiation, visible or otherwise, our concept of light has ultimately expanded to encompass a very wide range of wavelengths, from Ångströms to millimeters. These days, electromagnetic waves outside this range can also be referred to merely as *light*. We can now readily produce light spanning the entire spectrum, from the shortest wavelength light, called X-rays, to the longest, referred to as infrared (IR). Roughly speaking, light in the former of these wavelength ranges is usually considered to consist of particles with momenta $p = h\nu/c$, where h is Plank's constant, ν is the frequency, and c the speed of light, because scattering is most easily described in terms of momentum. The wavelengths ($\lambda = c/\nu$) of Vis and IR light may be vastly greater than the object to be observed if the object is very small, such as a nanometer-scale molecule. In such cases, the light interacts with charged particles in the object—protons in the nuclei and electrons—as a homogenous electric (and sometimes magnetic) field. Light in the Vis to ultraviolet (UV) range is often produced by electronic transitions, while IR is chiefly associated with spring-and-ball-like vibrations of molecules and lattice vibrations in crystals. Across the spectrum, light follows the wave-particle duality relation, $p = h/\lambda$, and the law of conservation of momentum holds for scattering between the object and

---

K. Shudo (✉)
Faculty of Engineering, Yokohama National University, Tokiwadai 79-5, 240-8501
Hodogaya-ku Yokohama, Japan
e-mail: ken1@ynu.ac.jp

K. Shudo et al. (eds.), *Frontiers in Optical Methods,*
Springer Series in Optical Sciences 180, DOI: 10.1007/978-3-642-40594-5_1,
© Springer-Verlag Berlin Heidelberg 2014

1

the light. Developments in the techniques of light generation and detection allow a very wide variety of wavelengths to be utilized. The reader's interests need not be limited to the light itself, but may include the control and modulation of light and related phenomena.

In a material, electric and magnetic fields (D and B, respectively) are induced by finite densities of electric and magnetic polarization (P and P_m, respectively), in a medium under external electric and magnetic fields (E and H). The following relations define these quantities:

$$D \equiv \epsilon E = \epsilon_0 E + P \tag{1.1}$$
$$B \equiv \mu H = \mu_0 H + P_m \tag{1.2}$$

where ϵ is the dielectricity and μ the magnetic permeability of the material. Note the suffix zero denotes the vacuum value, implying $P = 0$ and $P_m = 0$. The electric polarization plays an important role in the photoabsorption and reflection of light by the material.

All these physical quantities are governed by **Maxwell's equations**:

$$\nabla \cdot D = \rho \tag{1.3}$$
$$\nabla \cdot B = 0 \tag{1.4}$$
$$\nabla \times E = -\frac{\partial}{\partial t} B \tag{1.5}$$
$$\nabla \times H = \frac{\partial}{\partial t} D - J \tag{1.6}$$

which are mediated by the electric charge density ρ and the electric current density J. Two wave equations, derived from (1.3) through (1.6),

$$\nabla^2 E = c' \frac{\partial^2}{\partial t^2} E \tag{1.7}$$
$$\nabla^2 B = c' \frac{\partial^2}{\partial t^2} B \tag{1.8}$$

give simple independent solutions of the electric and magnetic fields at the point r and time t

$$E \equiv E(r, t) = E_0 e^{i(k \cdot r - \omega t)} \tag{1.9}$$
$$B \equiv B(r, t) = B_0 e^{i(k \cdot r - \omega t)}. \tag{1.10}$$

These fields propagate with velocity $c' = c/n$ through a material with refractive index $n = \sqrt{\frac{\epsilon \mu}{\epsilon_0 \mu_0}}$ (note that the velocity of light in a vacuum is $c = \sqrt{\frac{1}{\epsilon_0 \mu_0}}$). The optical responses of materials — how waves of light propagate through it — are formulated in terms of the modulation of the above wave equations. The assumptions that the materials are non-magnetic ($\mu \simeq \mu_0$) and electrically neutral ($\rho = 0$) are sufficient to discuss the following issues.

In a material medium, light is absorbed. This can be expressed mathematically as an expansion of the definition of the refractive index . Rather than using a purely real number n to describe the refractive index, the complex number $\tilde{n} = n + i\kappa$ can be introduced, where n and κ are real ($i^2 = -1$). Consequently, the wavevector $\tilde{k} = k' + ik''$, and the dielectricity $\tilde{\epsilon} = \epsilon_1 + i\epsilon_2$ of the corresponding wave equations are complex. The propagating wave at depth z can then be derived from (1.9) as

$$E = E_0 e^{i\omega(\frac{n}{c}z - t) - \frac{\omega\kappa}{c}z} \qquad (1.11)$$

and the intensity (density of energy of the electric field) is given by

$$I = I(z) = \frac{\epsilon}{2}|E|^2 = I_0 e^{-\alpha z} \qquad (1.12)$$

where $\alpha = 2\omega\kappa/c$ is called the absorption coefficient of the material with extinction coefficient κ. Thus, one irradiates a material of thickness d with light of wavelength λ, and the intensity of the light that emerges from the other side of the material is reduced by the exponential factor $e^{-\alpha d}$. Equivalently, α can be determined from the intensity I of light that passes through the material. Determining the wavelength (or energy) dependence of α is the most classic form of *spectroscopy*. This is directly related to the dielectricity, which determines transmission and reflection. However, these can be modulated by irradiation with high intensity light. This modulation, chiefly due to the generation of excited electronic states and crystal lattice vibrations, maintains a oscillatory phase. This type of modulation, referred to as *coherent*, is emphasized here, because most optical techniques utilize the interference of light. Contemporary applications of coherence are addressed in the following Chapters. The macroscopic quantities described above can be explained in terms of photon-electron and electron-lattice interactions. Thus, electronic and lattice behavior can be analyzed with appropriate light sources. As the size of the material will be smaller than the wavelength of the irradiated light in most cases, we must deal with these quantities within the framework of quantum mechanics of object material systems.

Consider the Schrödinger equation, with Hamiltonian $\hat{H}$, where we define $\hat{H}_0$ as the electronic Hamiltonian without electromagnetic fields due to the incident light,

$$i\hbar\frac{\partial}{\partial t}\psi = \hat{H}\psi. \qquad (1.13)$$

This can be rewritten as a secular equation $\hat{H}\psi = \mathcal{E}\psi$ in the case of stationary states. This has solutions indexed by their state j, represented by a wavefunction $\psi = \phi_j(\mathbf{r})$ with eigenvalue $\mathcal{E}_j$, corresponding to the energy of the electron. If one does not consider thermal effects at high temperatures, the highest energy states which are occupied by electrons, called the Fermi level E_F, is simply obtained from Pauli's exclusion principle. When the system is irradiated with light, the photoabsorption is described in terms of the probability of an electron in a ground state (ϕ_g with energy below the Fermi level) to make a transition into an excited state (ϕ_x above the Fermi level). To deal simply with the electromagnetic field of light, we use a simple electric-dipole approximation. Because magnetic fields do not do work on the electrons, they are not considered here, as spin and angular momenta cancel macroscopically in most non-magnetic ($\boldsymbol{P}_m = \boldsymbol{0}$, or $\mu \simeq \mu_0$) materials. Including the electric field $\boldsymbol{E}$ as a perturbation, the Hamiltonian becomes

$$\hat{H} = \hat{H}_0 + \hat{H}' \tag{1.14}$$

with the electric potential energy of electrons j at point $\boldsymbol{r}_j$ being given by

$$\hat{H}' = \sum_j -\boldsymbol{E}(\boldsymbol{r}) \cdot (-e)\boldsymbol{r}_j, \tag{1.15}$$

where $\boldsymbol{E}$ is the externally applied light. Note that $\boldsymbol{M} = \sum_j (-e)\boldsymbol{r}_j$ is the electric dipole moment of the system with each elemental charge $-e$. The wavelength of ultraviolet (UV) light with photon energy up to $\sim$10 eV is of the order of 100 nm. Wavelengths in the region of Vis to UV are much larger than the atomic scale, and the external field can be well described as a spatially uniform, but time-dependent, field.

Expanding the wavefunction in Schrödinger's equation in terms of the sates ϕ_j and their (time dependent) amplitudes $b_j(t)$, such that $\psi = \sum_j b_j(t)\phi_j$, the time evolution of the amplitudes are expressed as

$$i\hbar \frac{\partial}{\partial t} b_j(t) = \sum_l b_l(t) H'_{j,l} e^{i\omega_{j,l}t}, \tag{1.16}$$

with matrix elements

$$H'_{j,l} = \left\langle \phi_l | \hat{H}' | \phi_j \right\rangle. \tag{1.17}$$

An initial condition may be that the initial state of a perturbed wave function is $\psi = \phi_g$ at $t = 0$. If the basis set of wavefunctions ϕ_j is normalized, the coefficient

$$\frac{\partial}{\partial t} b_j(t) = \frac{i \boldsymbol{M} \cdot \boldsymbol{E}_0}{\hbar} \cos \omega_{j,g} t \, e^{i\omega t} \tag{1.18}$$

is easily obtained, where a difference of the eigenfrequencies $\omega_{j,g} = \frac{\mathcal{E}_j - \mathcal{E}_g}{\hbar}$ is used. An extinction term is added for a realistic description of the finite lifetime that the excited state generally has, and it becomes

$$\frac{\partial}{\partial t} b_j(t) = \frac{i \boldsymbol{M} \cdot \boldsymbol{E}_0}{\hbar} \cos \omega_{j,l} t \, e^{i\omega t} - \frac{\Gamma}{2} b_j(t) \tag{1.19}$$

which makes the probability $|b_j(t)|^2$ of a component j decay as $\propto e^{-\Gamma t}$. Finally, the focused component of the excited state x is described in terms of the coefficient

$$b_x(t) = \frac{i\boldsymbol{M} \cdot \boldsymbol{E_0}}{2\hbar} \left\{ \frac{e^{i(\omega+\omega_{x,g})t}}{\omega + \omega_{x,g} - i\frac{\Gamma}{2}} + \frac{e^{i(\omega-\omega_{x,g})t}}{\omega - \omega_{x,g} - i\frac{\Gamma}{2}} \right\}. \tag{1.20}$$

In isotropic media, the averaged product of $|\boldsymbol{M} \cdot \boldsymbol{E_0}|^2$ over all directions can be replaced by $\frac{1}{3}|\boldsymbol{M}|^2|\boldsymbol{E_0}|^2$. As the intensity of excited state formation is $|b_x|^2$, from (1.20), the averaged transition (photoabsorption) probability (frequency per unit time) in the medium due to the applied light of intensity I, is given by

$$B = \lim_{t \to \infty} \frac{|b_x(t)|^2}{t\,I} = \frac{\pi M^2}{3\epsilon_0 \hbar^2}, \tag{1.21}$$

which is referred to as Einstein's B-parameter. Essentially, this is the origin of what is called a spectrum: B plotted against energy (or frequency). This classic picture is still remarkably eloquent, being able to deal with the propagation of electromagnetic fields through materials. This dipole model is often applied to molecular systems, as an isolated molecule is spatially limited to the scale of nanometers. Eigenvalues of a molecule are discrete, being separated by an energy gap of the order of an eV, and corresponding electronic transitions resonantly occur in the Vis to near-UV region. Although the reason for the decay is yet to be clarified, spontaneous and artificial relaxation of excited electrons is now able to be observed, owing to the advent of ultrafast lasers with the ability to achieve femtosecond pulse durations. A contemporary discussion of topics in quantitative experiments is given in the ensuing chapters.

1.1.2 Electronic Bands

In solids, the scale of the system can be larger than the wavelength of Vis to near-UV light. To overcome this, infinite periodicity is assumed in order to solve the wave equation of electrons. A solution of the Schrödinger equation in an infinitely periodic potential is called a Bloch state, which forms electronic bands. Electrons with a momentum $\boldsymbol{p}$ are identified by a wave vector $\boldsymbol{k} = \boldsymbol{p}/\hbar$ shown in Fig. 1.1a. As electronic states in different Brillouin zones are identical, the states are identified with a value of $\boldsymbol{k}$ in the first Brillouin zone. Photoabsorption is an electronic transition between two bands. Figure 1.1(b) shows electronic bands, for example, of a Si crystal whose structure is face-centered cubic (f.c.c.). As the momentum of a photon is negligibly small compared with the momentum of Bloch electrons, the momentum of an electron can be considered to be constant during an electron-photon collision. Electronic transitions by photoabsorption process are thus shown as vertical arrows, as shown in Fig. 1.1b. In these cases, the dipole approximation takes another form. The perturbed Hamiltonian of a charged particle (with an electronic charge q), is [1],

$$\hat{H} = \frac{1}{2m}\left(\hat{\boldsymbol{p}} - q\boldsymbol{A}\right)^2 + \hat{V} \tag{1.22}$$

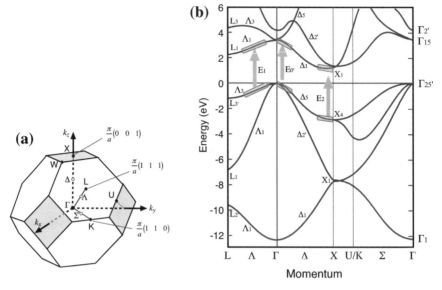

Fig. 1.1 **a** Electronic momenta in reciprocal space $k = (k_x, k_y, k_z)$. The polygon defines the first Brillouin zone of a f.c.c. lattice with lattice constant a. Symbols of high symmetry points, after [2], are marked at various points. **b** Band structure of a silicon crystal along the momentum $p = \hbar k$, following a path between symmetry points in the Brillouin zone. The vertical axis is the electronic band energy, defined to be zero with respect to the valence band maximum. Data is taken from [3]. Vertical arrows indicate typical photo-transitions

in which the vector potential is A, the momentum operator $\hat{p} = \frac{\hbar}{i}\nabla$, and the potential due to electrons and nucleus is $\hat{V}$. This form of the interaction is called the minimal interaction representation, or minimal coupling, of the electromagnetic field. Equations (1.14) and (1.15) describes one of the converted forms of this formula (See Appendix A).

Some people prefer (1.22) because this form of Hamiltonian explicitly contains both electric and magnetic fields, and also because this formula is in a relativistically invariant form [4]. From the point of view of optical transitions, this is particularly easy to use

$$\hat{H} = \hat{H}_0 + \hat{H}' \tag{1.23}$$

$$\hat{H}_0 = -\frac{\hbar^2}{2m}\nabla^2 + \hat{V} \tag{1.24}$$

$$\hat{H}' = \frac{q}{2m}\left(A \cdot \hat{p} + \hat{p} \cdot A\right) \tag{1.25}$$

$$\simeq -\frac{e}{m}\left(A \cdot \hat{p}\right), \tag{1.26}$$

Fig. 1.2 The band structure of a free electron in a f.c.c. crystal, as a function of wave vector **k**, in the directions of Λ and Δ. In the first Brillouin zone, they are the same directions as in Fig. 1.1. Along the Δ direction (*gray areas*), the adjacent second zone is also drawn, to make the parabolic nature of the dispersion clear

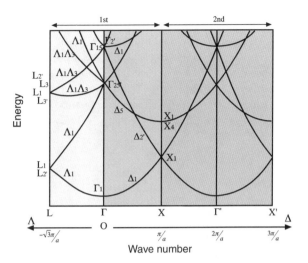

where $q = -e$ and the Coulomb gauge ($\nabla \cdot A = 0$) has been used. The long wavelength limit ($A \cdot \nabla = \nabla \cdot A$) of the last approximation is taken. A quantity $|A|^2$ is generally omitted as it is much smaller than $A \cdot \hat{p} + \hat{p} \cdot A$. The term $|A|^2$ gives the energy density of the electromagnetic field of the irradiated light, and is taken to be a constant value if the photoabsorption is weak. When the incident light is very strong, or the medium material has particularly strong photo-absorption, this term should be considered because second order processes may occur. One sets **k** parallel to the direction of propagation of the electromagnetic field, then the Coulomb gauge enforces that the transverse wave (**A** perpendicular to the propagation direction) is

$$A(r,t) = A_0 e^{i(k \cdot r - \omega t)} \tag{1.27}$$

$$\nabla \cdot A = A_0 \cdot k e^{i(k \cdot r - \omega t)} = 0 \tag{1.28}$$

for all (r, t). Thus,

$$E = -\frac{\partial}{\partial t} A(r, t) = i\omega A_0 e^{i(k \cdot r - \omega t)} \tag{1.29}$$

$$B = \nabla \times A = ik \times A_0 e^{i(k \cdot r - \omega t)}. \tag{1.30}$$

This is a standard expression for a light wave in the semiclassical treatment of radiation [5].

The result from reciprocal (i.e., **k**-) space with (1.26) and that from real (**r**-represented) space in (1.15) must be the same (See Appendix B). To analyze interband transitions, it is most easy to handle the transition perturbation in **k**-space than in **r**-space, because the dispersion is plotted as a function of a good quantum number (momentum or wavenumber). A parabolic band structure of free electrons (described in terms of plane waves) in a f.c.c. lattice is shown in Fig. 1.2. This is very similar

to that of actual Si bands in Fig. 1.1. This indicates that the real band may be well
approximated by plane wave states. A slight difference between the two, and one
which is very important to us here, is the band gap near E_F. An actual wavefunction
ψ consists of a plane wave, with wavenumber k, as well as higher order components
with the wave number shifted by G; see duplicity between regions $\Gamma - X$ and $\Gamma' - X'$,
for example. The shift G is a reciprocal lattice vector, defined as $\frac{2\pi}{a}(n_x, n_y, n_z)$ with
$n_{x,y,z} = \pm 1, \pm 2, ...$ The resultant wavefunction of the electron is

$$\phi_k(r) = \sum_{n_x,n_y,n_z} C_{n_x,n_y,n_z}(k) e^{i(k-G)\cdot r}, \qquad (1.31)$$

where $C_{0,0,0} \simeq 1$. Note that a set of integers (n_x, n_y, n_z) describes the three-
dimensional Brillouin zone. These higher order contributions, which deviate from
the free electron state, are generated by the perturbing potential V, which originate
from the underlying lattice and those electrons other than the nearly free electron
being described. After performing a Fourier expansion of the potential, given by

$$V(r) = \sum_{n_x,n_y,n_z} V_{n_x,n_y,n_z} e^{iG\cdot r}, \qquad (1.32)$$

a potential component V_{n_x,n_y,n_z} primarily affects the coefficient C_{n_x,n_y,n_z} of the
perturbed wavefunction [6]. A realistic band, as shown in Fig. 1.1, can be obtained
with only a few potential components [7] in a simple and highly symmetric crystal
structure such as silicon. Thus, the transition matrix element of the perturbation term,
(1.26), is

$$\langle \phi_j | \hat{H} | \phi_g \rangle = -\frac{e}{m} \langle \phi_j | A \cdot \hat{p} | \phi_g \rangle \qquad (1.33)$$

$$= -\frac{e}{m} A \cdot \sum_{n_x,n_y,n_z \neq 0} \langle \phi_j | \hat{p} | \phi_g \rangle \qquad (1.34)$$

$$\simeq -\frac{e\hbar}{m} A \cdot \sum_{\text{few } n_x,n_y,n_z} (k + G) \langle \phi_j | \phi_g \rangle. \qquad (1.35)$$

Remember that both ϕ_j and ϕ_g have the same wave vector k, because momentum
conservation restricts the transitions vertical as the arrows in Fig. 1.1. Sets of coefficients $\{C_{n_x,n_y,n_z}\}$ in (1.35), differ slightly from (1.31), determine the wavefunctions.
Therefore, the term $\langle \phi_j | \phi_g \rangle$ can easily be obtained, and is given by

$$\left\langle e^{i(\frac{p}{\hbar}-G)} | e^{i(\frac{p}{\hbar}-G)} \right\rangle = \frac{1}{2\pi} \sum_{\text{the few } n_x,n_y,n_z} C^*_{n_x,n_y,n_z}(j) C_{n_x,n_y,n_z}(g) \qquad (1.36)$$

Thus, the transition probability can be obtained and it should be proportional to the
imaginary part of the dielectricity (ϵ_2). The reflectivity is also calculated from ϵ_2,

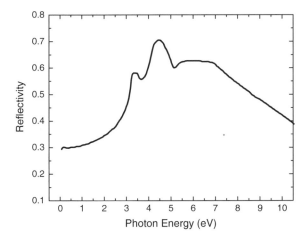

Fig. 1.3 Reflectivity spectrum of a crystalline Si surface. Data is taken from [8]

and agrees well with experiments [10, 8] (see the next section). An experimentally observed reflectivity of Si is shown in Fig. 1.3. There are peaks in the reflectivity at approximately 3.2 and 4.3 eV. The lower peak has two components, E_0 and E_1, while the higher peak contains a single component, denoted E_2 [11]. Corresponding electronic transitions are marked as arrows in Fig. 1.1, which connect two nearly parallel bands [12]. The shape of the band, and the peaks around E_1 and E_2, are common to materials in the zinc-blend family [3]. In these regions, photoabsorption occurs with high probability [8].

1.2 Surface Systems

As synchrotron radiation (SR) can be utilized these days, we are able to measure the precise optical responses in the UV to X-ray region of the spectrum. With the light in these region, photoemission spectroscopy is available to obtain the density of state (DOS) [13]. Generally, in interpreting photoemission spectra, we must consider the DOS of the final state as well as the initial state [14]. With the tunable light source (i.e., SR, laser, etc.), we can choose the wavelength of the irradiated light to reveal the DOS. In the experimental measurements of optical responses and electronic states of materials, features specific to surface system should be considered generally as any probe or signal have to pass the interface/surface of the medium inevitably. On the other hand, electronic states can now be easily determined via numerical calculations within a density functional theory (DFT) scheme, and using pseudopotentials is effective with a generalized gradient approximation (GGA) so far as isolated or inner-core electrons can be replaced with the potentials [15, 16]. This method has been successfully applied to semiconductors [17], as well as transition and nobel metal surfaces [18]. To obtain the density of states of a Si crystal, as shown in Fig. 1.4a, the reciprocal space was divided into an $8 \times 8 \times 8$ mesh, for example. The calculated spectra reproduce quite well the experimentally obtained spectrum for energies below the Fermi energy, E_F (Fig. 1.4c). However, there is a shoulder in the actual experimental data of clean Si. This is attributed to surface states, originating from the dangling bonds (unpaired electrons at the surface, pointing out towards the vacuum) [19]. To represent the surface of a system having infinite periodicity, we can employ a slab-like geometry (Fig. 1.5). A DOS spectrum of the slab calculated (with a $6 \times 6 \times 1$ k-space mesh) is shown in Fig. 1.4b, and surface states below and above E_F are observed (indicated by arrows in the figure). Note that even a sparse k-space mesh is sufficient to observe the surface states [20]. The escape depth of electrons (mean-free path without scattering) has a minimum value of around $\simeq 10$ Å when the electron energy is a few tens of eVs [21, 22]. The intensity of the calculated surface state response is consistently similar to that observed in the experimental spectra.

When a Si crystal is compressed or expanded, the electronic states change. In the chapter on the topic of high-pressures, a description is given of how the optical response of compressed materials is measured as temperature is varied. In Fig. 1.6, the calculated electronic states of a crystal are shown, where the lattice size of the crystal is changed isotropically. The total energy of a unit cell in the lattice, Fig. 1.6a, shows a parabolic well, which corresponds to the harmonic oscillator-like lateral breathing mode. This compression/expansion changes the electronic structure slightly, as shown in Fig. 1.6b plotted in terms of DOS. A shoulder at $E_F - 0.7$ and $E_F + 0.5$ eV

Fig. 1.4 The calculated density of states (DOS) with respect to valence band maximum of a silicon crystal. *a* Three-dimensional crystal with diamond lattice periodicity, representing a perfect bulk. *b* Two-dimensional structure of 6 layers of Si with a surface of Si dimers. The other surface was terminated with hydrogen. Surface states, which originate from dangling bonds at the dimers, appear as a shoulders in the DOS. *c* Photoemission spectroscopy using photons with energy $h\nu = 25$ eV. Data is taken from [9]

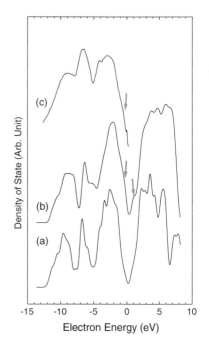

Fig. 1.5 An example of a slab, representing a surface. The structure is a p(2 × 2) structure of the Si(001) surface, containing two Si dimers buckled in different directions (toward $+x$ and $-x$) [19]. To terminate the dangling bond on the other side, extra hydrogen atoms are added (bottom of the slab)

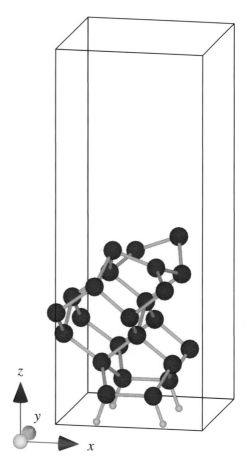

shifts slightly due to this deformation. The bulk-derived state at $E_F \pm 1.5$ eV on the other hand, changes considerably, the shifts in the large peaks between -2.4 to -2.0 eV and between $+2.1$ and ~ 3 eV are particularly noteworthy. These correspond to the shift of the E_1 and E_2 peaks [12]. Thus, the transitions near E_0, E_1, and E_2 are all very sensitive to expansion (or other distortions) of the lattice. The reflection rate of photons with energy of a few eV changes when a phonon causes the lattice size to vibrate. With ultrafast pulsed lasers, real-time oscillations in the reflectance due to the compression/expansion of the lattice can be measured [23]. In the following chapters, such ultrafast measurements are introduced. On the other hand, it is very difficult to detect the surface states as compared with the bulk. However, the change of the surface states can be detected when the reflectance is precisely measured. In the next section, the particular case of the sensitivity of the reflectivity to the surface states is explained. Experimental results, by means of an optical modulation method, are addressed in another Chapter, and are also described in detail.

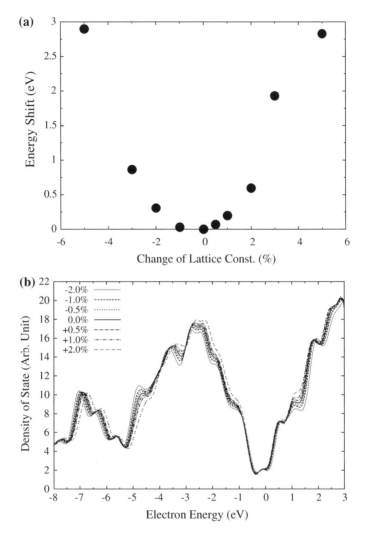

Fig. 1.6 **a** Calculated values of the relative energy per unit cell and **b** the density of states with respect to the Fermi level E_F of the slab model in Fig. 1.5 when the lattice constant is varied

1.2.1 Surface Reflectance

The optical response in terms of reflectivity has been widely used to measure thin films. Ellipsometry has been often applied to determine the thickness of a film on a surface if the dielectric constants of the substrate- and the surface-materials are known [24]. In recent years, optical spectroscopic methods, such as **surface differential reflectivity (SDR)** and **reflectance difference spectroscopy (RDS)** (sometimes referred to **reflectance asymmetry (RA)**), have been developed as powerful tools to investigate the electronic properties of semiconductor surfaces and interfaces, even at the atomic scale [25–27]. The principle advantage of these methods is their ability to observe processes on the surfaces nondestructively in real time, even in high-pressure reactive gasses. However, optical spectra alone are not sufficient to provide an understanding of the surface structures and the electronic states that determine the optical responses. Theoretical analysis of the surface electronic states is necessary to interpret the optical spectra in terms of surface structures.

To assign the optical spectra to surface states, we calculated the electronic states and optical transitions of the Si(111) surface using small clusters. Although the Si(001) surface is chiefly used in silicon devices, Si(111) is more frequently used in research at the atomic scale, because of the variety of reactions due to its reconstructed structure. Thus, we report here the calculated optical spectra of the Si(111)-7×7 dimer–adatom–stacking-fault (DAS) surface [28]. The adsorption process of chlorine at the Si(111) surface can be divided into two stages. First, a chlorine atom reacts with silicon at adatom (AD) sites to form monochloride, to terminate the dangling-bond (DB) [29, 30]. When the adsorption density becomes almost equal to that of the adatoms, di- and trichloride begin to form [31, 32]. When these polychlorides are produced, one of the three adatom back-bonds (ADBB) is broken. Then, a Cl atom is bonded to the adatom, leaving a rest-atom dangling-bond (RADB) [33, 34]. We have calculated SDR spectra $\Delta R/R$ of the mono- and dichloride Cl/Si(111) surfaces, and analyzed the electronic states associated with features in the spectra.

1.2.2 Calculation

While density functional theory (DFT) deals only with the ground states, time-dependent density functional theory (TD-DFT) can deal with excited states as well as the ground states [35]. TD-DFT calculations can be performed on relatively small computers, such as desktop workstations. The calculation of optical spectra using TD-DFT has been reported not only for molecules [36], but also for solids [37], and TD-DFT allows the precise evaluation of the excited states of molecules. Clusters which contain 13 silicon atoms are sufficient for the present purpose, because Ricca et al. used clusters of this size to calculate the Si-Cl frequency using DFT, and the calculated value was in good agreement with the experimental value [38]. Therefore, we deem that this method should also be applicable to silicon surfaces. We calculated the electronic states and the optical transitions with the widely used *ab initio* calculation program Gaussian 03/09 [39].

The compositions of the clusters used to represent the T_4 site of Si(111) are $Si_{21}H_{27}$ for the clean surface, and $Si_{21}H_{27}Cl$ and $Si_{21}H_{27}Cl_2$ for chlorine-adsorbed surfaces. The last two represent monochloride and dichloride generated from the adatom, respectively. The geometric structures of the clusters used for the calculation are shown in Fig. 1.7. In order to eliminate all the dangling-bonds except at the Si adatom, all silicon atoms except the adatom are terminated by hydrogen atoms. In the calculation of the photo-transition energy of the clusters, the hybrid TD-DFT

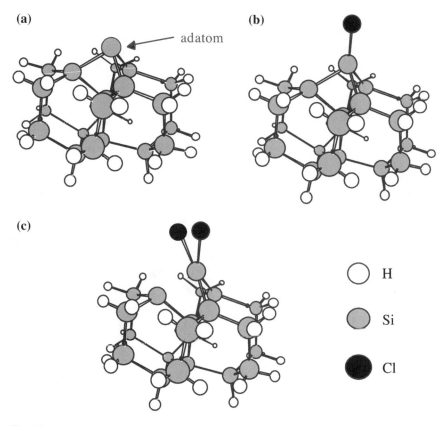

Fig. 1.7 Geometric structures of the clusters used to represent the Cl/Si(111) surface: **a** $Si_{21}H_{27}$, representing the clean surface; **b** $Si_{21}H_{27}Cl$ and **c** $Si_{21}H_{27}Cl_2$, representing chlorine-adsorbed surfaces

method was employed, because it gives the best agreement with experiment [40]. Furthermore, we selected the Becke 3-parameter hybrid functional with Lee, Yang, and Parr correlation (B3LYP), as the preferred tool [41, 42]. First, the geometry of the Si(111) clusters was energetically optimized at the 6-31G level. After the optimization, the calculations of electronic excitation were carried out with the 6-31G* basis set. The distribution of the wave-function relevant to the photo-transitions was examined with the visualization program Molden [43].

From the results for electronic excitations and oscillator strengths, optical transition probabilities were obtained. The oscillator strength f can be expressed in terms of transition probability μ as follows:

$$f_{ij} = \frac{2m\omega}{3e^2\hbar}|\mu_{ij}|^2 \qquad (1.37)$$

in which i and j represent the initial and final states, respectively, and

$$\mu_{ij} = \frac{\langle \phi_j | \boldsymbol{e} \cdot \boldsymbol{p} | \phi_i \rangle}{|\boldsymbol{e}|}. \tag{1.38}$$

where $\boldsymbol{e}$ is the direction of the electric field.

The imaginary part, ϵ_2, of the dielectric function for frequency ω was obtained with the following expression [44, 45].

$$\epsilon_2(\omega) = \frac{4\pi^2 e^2}{m^2 \omega^2 A} \sum_{i,j} \delta[E_i - E_j - h\omega] \times |\mu_{ij}|^2, \tag{1.39}$$

where A is the volume of the cluster.

In the numerical calculation of (1.2), the δ-function is substituted by a normalized Gaussian function whose standard deviation is set to $0.1 \sim 0.15$ eV, because the photo-excitation has a band width of about this value in the bulk [46, 47] and at the surface [48]. The imaginary part is transformed into the real part through the Kramers-Kronig relation. Electronic transitions were calculated up to 6 eV, but this is not enough to calculate ϵ_2 and SDR spectra. To avoid this limit, ϵ_2 should be extrapolated to connect it to the imaginary part of the dielectric function in the bulk state [49]. Therefore, we multiplied the calculated ϵ_2 by a factor so that the imaginary part of the dielectric function in the bulk state is linked smoothly at energies greater than 3.1 eV for the clean surface and 4.9 eV for the Cl-adsorbed surface. Even if the factor is changed, the form of the SDR spectra does not qualitatively change. This factor corresponds to the cluster volume, A in (1.2). The index of refraction $n(\omega)$, the extinction coefficient $\kappa(\omega)$ and reflectance $R(\omega)$ can be derived using the complex dielectric function for the frequency ω. In the case of normal incidence, R can be expressed with Fresnel's formula.

Although s-polarized and p-polarized SDR spectra generally differ in intensity, the shapes of the spectra are very similar to each other in the case of Si(111) [26]. Consequently, we did not consider the polarization of incident light, and the calculation covered only normal incidence. By means of these procedures, an SDR spectrum $\Delta R/R$ can be calculated from the reflectances of the clean surface R_{clean} and the adsorbed surface $R_{\text{Cl-adsorbed}}$.

$$\frac{\Delta R}{R} = \frac{R_{\text{Cl-adsorbed}} - R_{\text{clean}}}{R_{\text{clean}}} \tag{1.40}$$

Next, we calculated the optical transitions of the cluster representing dichloride, shown in Fig. 1.7c, in which one of the ADBB is broken and terminated with a chlorine atom, and the rest-atom dangling-bond appears. The SDR spectrum for this back-bond breaking was calculated.

1.2.3 Example and Interpretation

Dangling-Bond termination

Calculated SDR spectra for monochloride formation during Cl-adsorption on Si(111) are shown in Fig. 1.8. The basis set effect between 6-31G and 6-31G* is negligible. Up to 3.2 eV, the spectrum computed with the 3-21G basis set resembles the other spectra. It is sufficient to analyze surface-to-surface transitions up to 3.2 eV, because electronic excitations of silicon above 3 eV are mainly contributed to by the bulk state. The positions of the peaks obtained with the 3-21G basis set are very close to those calculated with other basis sets, but the intensity is greater with 3-21G than with others. Although it is not yet clear why the intensity is sensitive to the basis set, the 3-21G basis set seems adequate to assign the spectral profile.

The experimental Cl/Si(111) SDR spectrum is shown in Fig. 1.9 [50]. The observed spectrum can be decomposed into two component spectra, S_A (a) and S_B. S_A corresponds to the negative peak at 1.55 eV in the calculated SDR spectra (b), similar to the assignment of the negative peak (c) for H/Si(111) to the calculated peak (d) at 1.4 eV [51]. Thus, the calculated feature at 1.55 eV (transition from the ADBB state to the antibonding ADBB state) is considered to be the origin of the main peak. The negative minimum in (b) and (d) at $2.1 \sim 2.2$ eV is due to the transition from an ADBB state to an antibonding ADBB state, although no peak was apparent experimentally. The calculated minimum at 2.8 eV, again not visible experimentally, is due to transitions from ADDB or ADBB to antibonding ADBB states and from bulk to antibonding ADBB states. As the energy becomes larger, bulk-to-bulk transitions become predominant, and above 2.8 eV, electronic transitions occur

Fig. 1.8 Cl/Si(111) SDR spectra calculated with *a* B3LYP/3-21G,*b* B3LYP/6-31G, and *c* B3LYP/6-31G* basis sets

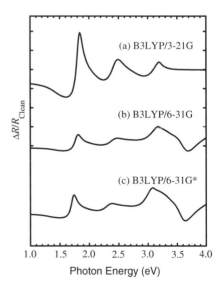

Fig. 1.9 SDR spectrum of S_A experimentally obtained for the monochloride Si(111) a and the spectrum obtained by calculation b. The experimental c and calculated d spectra of monohydride Si(111) [26] are plotted for comparison

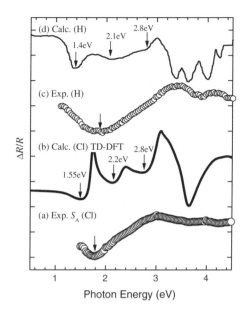

mainly inside the bulk. Thus, the assignment of spectral features of the Cl/Si(111) surface is possible with this method. The features are quite similar to those obtained in the case of hydrogen.

Back-Bond breakage

We next focused on the process of ADBB breakage, namely the process of dichloride formation, with the cluster having with the newly emerged dangling-bond at the rest-atom. In the model, the dangling-bond at AD is terminated with the Cl atom, while a dangling-bond remains at RA. The inclination of adatom dichloride towards the RADB, set to the initial geometry shown in Fig. 1.7c, was almost removed after the energetic optimization, in accordance with a microscopic estimation of Cl density [52].

A spectrum was calculated through the same approach as in the case of dangling-bond termination, and is shown in Fig. 1.10 together with the observed spectrum, S_B. The ΔR is defined here as $\Delta R \equiv R_{SiCl_2} - R_{SiCl}$, as S_B represents the difference of reflectance between mono- and dichloride. The results for dihydride formation obtained by experiment (c) and by means of huge cluster calculations (d) for hydrogen adsorption [26] are also plotted.

All the spectra in Fig. 1.10 contain a negative peak at around 2.6 ∼3.0 eV. This peak can be assigned to the loss of the transition at AD, owing to the missing BB. This assignment turned out to be commonly applicable to ADBB breaking by halogen and hydrogen. Thus, the TD-DFT method with small clusters is capable of interpreting

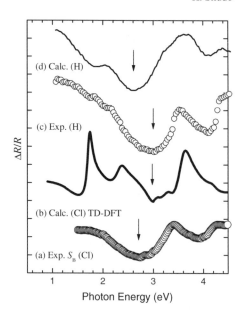

Fig. 1.10 a SDR spectrum of S_B experimentally obtained for the dichloride Si(111) and **b** the same spectrum obtained by a calculation. Experimental **c** and calculated **d** spectra for the dihydride [51] are also shown, for comparison

(d) Calc. (H)

(c) Exp. (H)

(b) Calc. (Cl) TD-DFT

(a) Exp. S_B (Cl)

$\Delta R/R$

Photon Energy (eV)

the SDR spectra for adsorption of elements with large electron affinity, such as chlorine. However, the spectrum (b) has some features below ∼2 eV. These may be caused by a perturbation of the remaining ADBB by the additionally adsorbed Cl atom, possibly through charge transfer from Cl to AD, as suggested by a previous calculation [53]. Although some transitions from/to the emerged DB at the rest-atom are from 2.6 eV and above, the effect of this kind of DB is still unclear. Further analysis will be required, however it is clear that the surface electronic transition is strongly affected by the surface electronic structure. Reflectance spectroscopy is dominated by the transition, and the surface structure, in terms of adsorbates and electronic states, can be elucidated with spectroscopic information.

1.3 Ultrafast Measurement

1.3.1 Monolayer Film of Organic Molecule

Among organic molecules, the thiol family is known to easily form a self-assembled monolayer (SAM) film with a bond at the sulfur site, even to noble metal surfaces. A benzentiol (BT) molecule contains the most fundamental aromatic functional that offers a variety of applications [54]. As molecular vibrations of such SAMs are fingerprints of adsorbed structures, we consider the dynamic motion of the molecules to elucidate molecule-substrate interactions in the ultrathin film. Ultrafast measurements of coherent molecular motion at the femtosecond time-scale are available with

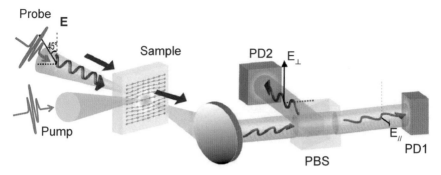

Fig. 1.11 Schematic illustration of the experimental setup for ultrafast transmission/absorption measurements

optical methods using ultrashort-pulsed laser systems. In this research, we measured the transient reflectance signal from a Au surface, on which the SAM of BT was adsorbed at the monolayer thickness. The resultant frequency domain spectra gave information on coherent motion of the adsorbed layer.

Strips of silicon (001) wafer were used as a substrate material. Prior to optical measurements the specimens were treated as follows: gold (Au) atoms were attached to the surface by sputter deposition to form Au films with thicknesses of a few Angstrom to several nm. The specimen was then soaked in an ethanol solution of BT (1 mM) for 24 hrs. On the Au surface, BT molecules were adsorbed at a monolayer thickness [55]. The surface morphology of Au was confirmed via atomic-force microscopy.

To obtain the ultrafast response, ultrashort pulses (7.5 fs) of a Ti:sapphire laser (800 nm) were used as a light source. For the liquid phase of the organic molecules, the transmission anisotropy was measured in real time by means of a pump-probe technique in electro-optical (EO) sampling configuration (Fig. 1.11). At the surface of the specimen, the optical response was measured in the reflectance configuration [57]. A pump beam was irradiated from the direction normal to the surface, and the beam was focused at the surface. A probe pulse delayed variably from the pump pulse was focused to the surface area where the pump was irradiated. The transmitted/reflected probe beam was divided into two polarized components orthogonal to each other. The difference of the intensities of the two components, measured with a pair of photodiodes, was detected electrically. This difference was taken as the sign of asymmetry in the optical reflectance due to surface deformation.

1.3.2 Experimental example

The transient signal of a BT monolayer film was plotted along the varied delay time as shown in Fig. 1.12. After an intense signal was observed initially (where 0 delay

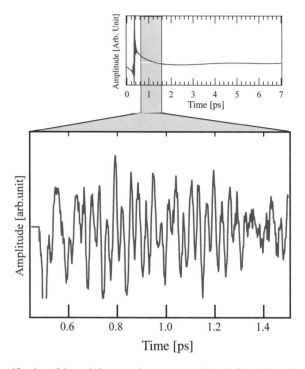

Fig. 1.12 Magnification of the real-time transient response, plotted after subtraction of a slow component. The upper panel is an observed transient response of the surface asymmetry in reflectance (EO sampling; see Text). The shadowed area in the upper panel corresponds to the main panel. Data taken from [56]

is at 0.37 ps), a slow decay was observed. After subtraction of the decay, a complex oscillatory signal remained (shown in the main panel). In the Fourier transformed spectra Fig. 1.13 from the residual signal, several peaks are found. In the figure, we also plotted peaks of vibrational components of liquid BT (> 98 % purity) obtained with conventional c.w.-Raman spectroscopy.

Most of the peaks in the Fourier spectra (gray in Fig. 1.13) correspond roughly to the vibrational peaks in the conventional spectra (black in Fig. 1.13). No peak due to the Si substrate was observed in the Fourier spectra, because we are able choose a polarization of the incident beam so as to excite the surface layer selectively, based on a selection rule associated with photoexcitation [23]. On the other hand, a peak originating from the crystalline Si (~16 THz) appeared as the strongest signal in the C. W. -Raman data (not shown). Thus, we can selectively distinguish the coherent motion of the two-dimensional structure of the BT monolayer, as the ultrashort pulses excite the SAM film impulsively. There are some discrepancies in the peak positions between the SAM film and the liquid phase. These vibration shifts can be explained in terms of immobilization of the molecule at the Au surface [54]. We calculated electronic and vibrational eigenvalues with a first-principle method, and the vibronic

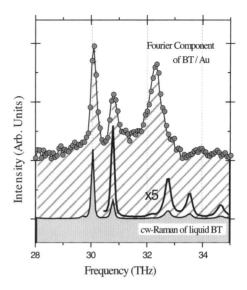

Fig. 1.13 Fourier components (hatched) of the subtracted signal (main panel in Fig. 1.12). Only a small range of the data, displaying typical peaks, is shown. In comparison, a conventional Raman spectra of liquid BT is shown (*gray*), which is magnified by a factor of five

modes of the BT adsorbed on the surface are in good agreement with the experimental results. The relative intensity of each mode varies when the thickness of Au deposition due to the sensitivity of Raman effect sensitive to the surface roughness [58]. This indicates the surface-enhanced Raman scattering (SERS) effect is active at the surface layer also in reflectance [57]. Morphology of the Au surface and alignment of the molecules seem to affect the ultrafast dynamics; the ultrafast pump-and-probe measurement enables us to discuss dynamic phenomena of real-time coherent motion in the monolayer thin films.

Using the pump-probe EO sampling technique, it is possible to selectively and sensitively measure the transient optical asymmetry originating from the vibration of organic molecules adsorbed on surfaces. To discuss the intermolecular, nonlinear, and low-dimensional lattice interactions in such molecular systems, real-time analysis gives new information [59] in terms of time-dependent intensity, phase, and beat frequency of each vibronic mode.

1.4 Summary

In this introductory chapter, the relation of the optical response of a material with its electronic states has been addressed. Next, the optical response of a surface system in the Vis to UV region was shown to be sensitive to an atomic-scale adsorbed material. The assignment of the electronic excitations for Cl/Si(111) was explained by means of calculations carried out in the TD-DFT regime, using small clusters. Finally, an ultrafast response of monolayer molecules adsorbed on a solid surface was discussed. This response was then related to the molecular vibronic modes, and the information about the real-time oscillations enables us to elucidate the vibronic phase and amplitude. This in turn gives information on atomic/molecular motion, including photo-induced excitation and relaxation processes.

1.5 Acknowledgment

The author is grateful to Prof. Masanobu Uchiyama (Riken, The Institute for Chemical and Physical Research) and Mr. M. Mori and Ms. E. Senga (Information Processing Center, YNU) for computational support. The calculations were partly performed on the Riken Integrated Cluster-of-Clusters (Riken) and the Supercomputer Facility at the Institute for Solid State Physics (the University of Tokyo), under advice concerning TD-DFT by Prof. Kaoru Ohno (YNU) and Dr. Soh Ishii (YNU). The author also thanks Mr. Tatsuya Momose for his assistance in performing the calculations and plotting the results, and Mr. Koshiro Doi for his assistance in analysing and drawing the ultrafast measurement data.

Appendix: A Long-Wavelength Approximation

Whenever a gauge transformation with a function χ,

$$\boldsymbol{A} \mapsto \boldsymbol{A}' = \boldsymbol{A} + \nabla\chi \tag{A.1}$$

$$U \mapsto U' = U - \frac{\partial}{\partial t}\chi, \tag{A.2}$$

is applied to a vector and scalar electromagnetic potential, $\boldsymbol{A}$ and U, simultaneously, the resultant fields $\boldsymbol{B}' = \nabla \times \boldsymbol{A}'$ and $E' = -\nabla U'$ are unchanged, such that

$$\boldsymbol{B} = \nabla \times \boldsymbol{A} \tag{A.3}$$

$$E = -\nabla U'. \tag{A.4}$$

To eliminate the magnetic field in (1.22), we take a function $\chi = -\boldsymbol{r} \cdot \boldsymbol{A}$. Then, in the case that the wavelength of the irradiating light is much larger than the size of the system, such as in the case of molecules (or crystalline domains where the electronic diffusion/drift is limited), the electric and magnetic fields can be considered to be constant. Then,

$$A' = A - \left((A \cdot \nabla)r + (r \cdot \nabla)A + A \times (\nabla \times r) + r \times (\nabla \times A)\right) \tag{A.5}$$

$$\simeq 0, \tag{A.6}$$

$$U' = U + \left((\frac{\partial}{\partial t}r) \cdot A + r \cdot \frac{\partial}{\partial t}A\right) \tag{A.7}$$

$$= U - r \cdot E, \tag{A.8}$$

where we note that r is independent of time, t. Finally, with A corresponding to the external light source from (A.3) and (1.5). $V = -eU$ being the electron-nucleus potential, the Hamiltonian (1.22), becomes the pair of equations, (1.14) and (1.15). This result is to be somewhat expected, because the magnetic field does no work on a charged particle unless the spin of the particle is considered from relativistic invariance. To analyse the magnetic effect of spin, the dipole approximation is not enough.

Appendix: B Minimal Interaction in r-Space

The uncertainty of quantum mechanical theory can be expressed as

$$\hat{p}_x x - x \hat{p}_x = \frac{\hbar}{i}. \tag{B.9}$$

Then,

$$\frac{1}{m}\frac{\hbar}{i}\hat{p}_x = \left(\hat{p}_x \frac{1}{2m}(x\hat{p}_x + \frac{\hbar}{i}) + Vx\right) - \left(\frac{1}{2m}(\hat{p}_x x - \frac{\hbar}{i})\hat{p}_x + xV\right) \tag{B.10}$$

$$= \left(\frac{1}{2m}\hat{p}_x(\hat{p}_x x) + Vx\right) - \left(\frac{1}{2m}(x\hat{p}_x)\hat{p}_x + xV\right) \tag{B.11}$$

$$= \left(\frac{1}{2m}\hat{p}_x^2 + V\right)x - x\left(\frac{1}{2m}\hat{p}_x^2 + V\right) \tag{B.12}$$

$$= \hat{H}x - x\hat{H}, \tag{B.13}$$

and the other components of p follow the same relation, giving

$$\frac{1}{m}\frac{\hbar}{i}\hat{p} = \hat{H}r - r\hat{H}. \tag{B.14}$$

Assuming an homogeneous electromagnetic field, A, the transition matrix element (1.17) in (1.16), using the perturbative term (1.26), becomes

$$\hat{H}'_{g,j} = \frac{e}{m}\langle\phi_j|\boldsymbol{A}\cdot\hat{\boldsymbol{p}}|\phi_g\rangle \tag{B.15}$$

$$= \frac{ie}{\hbar}\boldsymbol{A}\cdot\langle\phi_j|(\overleftarrow{H}\boldsymbol{r} - \boldsymbol{r}\overrightarrow{H})|\phi_g\rangle \tag{B.16}$$

$$= \frac{ie}{\hbar}\boldsymbol{A}\cdot\langle\phi_j|(\mathcal{E}_j\boldsymbol{r} - \boldsymbol{r}\mathcal{E}_g)|\phi_g\rangle \tag{B.17}$$

$$= ie\frac{\mathcal{E}_j - \mathcal{E}_g}{\hbar}\langle\phi_j|\boldsymbol{r}|\phi_g\rangle, \tag{B.18}$$

which leads to the same result as (1.18).

References

1. P.A.M. Dirac, *The Principles of Quantum Mechanics*, 4th edn. (Oxford University Press, Oxford, 1982)
2. J. Cornwell, *Group Theory and Eectronic Energy Bands in Solids* (John Wiley and Sons, New York, 1969)
3. J.R. Chelikowsky, M.L. Cohen, Phys. Rev. **B14**, 556–582 (1976)
4. J.D. Jackson, *Classical Electrodynamics*, 3rd edn. (John Wiley & Sons, New York, 1999)
5. Refer, for example, to Chapter 11 in it Quantum Mechanics, 3rd ed. by L. I. Sciff (Mc-Grawhill Inc., New York, 1955–1969).
6. M.L. Cohen, T.K. Bergstresser, Phys. Rev. **141**, 556 (1976)
7. E.O. Kane, Phys. Rev. **146**, 558 (1966)
8. H.R. Philipp, H. Ehrenreich, Phys. Rev. **129**, 1550–1560 (1963)
9. W.D. Grabman, D.E. Eastman, Phys. Rev. Lett. **29**, 1508–1512 (1972)
10. J.R. Chelikowski, M.L. Cohen, Phys. Rev. B **14**, 789 (1966)
11. R.R.L. Zucca, Y.R. Shen, Phys. Rev. B **1**, 2668 (1970)
12. M. Cardona, *Modulation Spectroscopy* (Academic Press, New York, 1969)
13. H. Ibach (ed.), Electron Spectroscopy for Surface Analysis (Topics in Current Physics) (Springer, Berlin, 1977 (Original ed.)).
14. W. Schattke, E.E. Krasovskii, R. Díez, P.M. Echenique, Phys. Rev. B **78**, 155314 (2008)
15. K. Ohno, K. Esfarjani, Y. Kawazoe, *Computational Materials Science: From Ab Initio to Monte Carlo Methods, Springer Series in Solid-State Sciences* (Springer, Berlin, 2000)
16. R.M. Martin, *Electronic Structure: Basic Theory and Practical Methods* (Cambridge University Press, Cambridge, 2008)
17. Y. Morikawa, Phys. Rev. B **63**, 033405 (2001)
18. T. Hayashi, Y. Morikawa, H. Nozoe, J. Chem. Phys. **114**, 7615 (2001)
19. A. Zangwill, *Physics at Surfaces* (Cambridge University Press, Cambridge, 1988)
20. A. Ramstad, G. Brocks, P.J. Kelly, Phys. Rev. B **51**, 14504–14523 (1995)
21. T.N. Rhodin, J.W. Gadzuk, Nature of the Surface Chemical Bond, ed. by T.N. Rhodin, G. Erti (North-Holland, Amsterdam, 1979), pp. 113–273.
22. D.R. Penn, Phys. Rev. B **13**, 5248–5254 (1976)
23. M. Hase, M. Kitajima, A.M. Constantinescu, H. Petek, Nature **426**, 51–54 (2003)
24. H. Fujiwara, *Spectroscopic Ellipsometry-Principles and Applications* (John Wiley & Sons, New York, 2007)
25. M. Palummo, O. Pulci, R. Del Sole, A. Marini, P. Hahn, W.G. Schmidt, F. Bechstedt, J. Phys.: Condens. Matter **16**, S4313 (2004)
26. C. Noguez, C. Beitia, W. Preyss, A.I. Shkrebtii, M. Roy, Y. Borensztein, R. Del Sole, Phys. Rev. Lett. **76**, 4923 (1996)

27. S. Ohno, T. Ochiai, M. Morimoto, T. Suzuki, K. Shudo, M. Tanaka, Jpn. J. Appl. Phys. **49**, 055702 (2009)
28. K. Takayanagi, Y. Tanoshiro, S. Takahashi, M. Takahashi, Surf. Sci. **164**, 367 (1985)
29. J.S. Villarrubia, J.J. Boland, Phys. Rev. Lett. **63**, 306 (1989)
30. J.J. Boland, J.S. Villarrubia, Phys. Rev. B. **41**, 9865 (1990)
31. R.D. Schnell, D. Rieger, A. Bogen, F.J. Himpsel, K. Wandelt, W. Steinmann, Phys. Rev. B **32**, 8057 (1985)
32. L.J. Whitman, S.A. Joyce, J.A. Yarmoff, F.R. McFeely, L.J. Terminello, Surf. Sci. **232**, 297 (1990)
33. K. Shudo, H. Washio, M. Tanaka, J. Chem. Phys. **119**, 13077–13082 (2003)
34. M. Tanaka, K. Shudo, M. Numata, Phys. Rev. B **73**, 115326 (2006)
35. J. Iwata, K. Yabana, Kotaibutsuri **39**, 771 (2004). (in Japanese)
36. Zachary H. Levine, Paul Soven, Phys. Rev. A. **29**, 625 (1984)
37. A. Marini, R. Del Sole, A. Rubio, Phys. Rev. Lett. **91**, 256402 (2003)
38. Alessandra Ricca, Charles B. Musgrave, Surf. Sci. **430**, 116 (1999)
39. URL: http://www.gaussian.com/
40. Mark E. Casida, Christine Jamorski, Kim C. Casida, Dennis R. Salahub, J. Chem. Phys. **108**, 4439 (1998)
41. Rüdiger Bauernschmitt, Reinhart Ahlrichs, Chem. Phys. Lett. **256**, 454 (1996)
42. R.E. Stratmann, G.E. Scuseria, J. Chem. Phys. **109**, 8218 (1998)
43. G. Schaftenaar, J.H. Noordik, J. Comput.-Aided Mol. Design, 14, 123–134 (2000) URL: http://www.cmbi.ru.nl/molden/
44. F. Manghi, R. Del Sole, A. Selloni, E. Molinari, Phys. Rev. B. **41**, 9935 (1990)
45. G. Allan, C. Delerue, Phys. Rev. B. **70**, 245321 (2004)
46. M. Murayama, T. Nakayama, Phys. Rev. B. **49**, 5737 (1994)
47. K. Shudo, T. Munakata, Phys. Rev. B. **63**, 125324 (2001)
48. K. Shudo, S. Takeda, T. Munakata, Phys. Rev. B. **65**, 075302 (2002)
49. Handbook of Optical Constants of Solid ed. by E.D. Palik (Academic Press Inc., London, 1985).
50. M. Tanaka, T. Shirao, T. Sasaki, K. Shudo, H. Washio, N. Kaneko, J. Vac. Sci. Technol. A **20**, 1358 (2002)
51. C. Noguez, A.I. Shkrebtii, R. Del Sole, Surf. Sci. **331–333**, 1349 (1995)
52. Y. Owa, K. Shudo, K. Koma, T. Iida, S. Ohno, M. Tanaka, J. Phys.: Condens. Matt. 18, 5895–5903 (2006).
53. P.V. Smith, P.L. Cao, J. Phys.: Condens. Matt. 7, 7125–7139 (1995).
54. R.C. Price, R.L. Whetten, J. Am. Chem. Soc. **2005**, 13750–13751 (2005)
55. H. Kang, T. Park, I. Choi, Y. Lee, E. Ito, M. Hara, J. Noh, Ultramicroscopy **109**, 1011–1014 (2009)
56. K. Shudo, K. Doi, I. Katayama, M. Kitajima, J. Takeda, *Ultrafast Phenomena XVIII, 05013* (EDP Sciences, Les Ulis Cedex, 2012)
57. I. Katayama, S. Koga, K. Shudo, J. Takeda, T. Shimada, A. Kubo, S. Hishita, D. Fujita, M. Kitajima, Nano Lett. **11**, 2648–2654 (2011)
58. M. Shindo, T. Sawada, K. Doi, K. Mukai, K. Shudo, J. of Phys.: Conference Series 441, 012044 (2013).
59. O.V. Misochko, M. Hase, M. Kitajima, Phys. Solid St. **46**, 1741–1749 (2004)

Part I
Reflectance Spectroscopy

Chapter 2
Real-Time Analysis of Initial Oxidation Process on Si(001) by Means of Surface Differential Reflectance Spectroscopy and Reflectance Difference Spectroscopy

Shin-ya Ohno, Ken-ichi Shudo and Masatoshi Tanaka

2.1 Introduction

In situ non-destructive measurements using optical methods are widely used to fabricate nanodevices and to elucidate catalytic phenomena. Here we introduce two types of optical method, namely, **surface differential reflectance (SDR)** spectroscopy and **reflectance difference spectroscopy (RDS)**. The SDR signal is obtained from the relative difference in reflectance of the clean and adsorbate-covered surface, while RDS probes the difference between the reflection coefficients for two mutually perpendicular polarizations. These are both linear optical techniques, like **spectroscopic ellipsometry (SE)**. SE measures the whole light reflected from thin films within an escape depth over 10 nm. Therefore, the surface sensitivity of SE has certain limitations, although the resolution in determination of the film thickness is very high (~0.01 nm). Both SDR and RDS techniques have the advantage of high surface sensitivity, because the bulk signal is efficiently eliminated, enabling us to analyze surface structure and reactions on the atomic scale.

The development of SDR and RDS was motivated in the 1980s by the demand for in situ real-time measurement tools which could be applied to the high-pressure processes in **molecular beam epitaxy (MBE)** and **metalorganic vapor phase epitaxy (MOVPE)** [1, 2]. The high surface sensitivity of these methods allowed them to be applied not only to semiconductor surfaces, but also to metal surfaces and solid–liquid interfaces [3, 4].

In the 1990s, it became possible to follow the formation of specific bonds, as well as adosrption/desorption at specific sites with the aid of first-principles calculations [5–7]. These studies demonstrated that additional information on concomitant chemical reactions occurring at different sites can be obtained if these techniques are employed in combination with conventional surface analysis tools.

S. Ohno (✉) · K. Shudo · M. Tanaka
Department of Physics, Faculty of Engineering, Yokohama National University, Yokohama, Japan
e-mail: sohno@ynu.ac.jp

K. Shudo et al. (eds.), *Frontiers in Optical Methods*,
Springer Series in Optical Sciences 180, DOI: 10.1007/978-3-642-40594-5_2,
© Springer-Verlag Berlin Heidelberg 2014

Recently, intensive studies have been performed on real-time analysis of surface chemical reaction processes by means of dispersive **near-edge X-ray absorption spectroscopy (NEXAFS)** [8] and rapid **X-ray photoelectron spectroscopy (XPS)** [9]. In the case of SDR and RDS, it is difficult to identify chemical species solely from the spectroscopic analysis. However, temporal resolution of the order of ~ms is possible [10], which, at present, is even better than can be obtained with the above techniques. The advantage of the optical methods is their suitability for real-time measurement under high-pressure gas exposure without loss of sensitivity. In addition, the use of a synchrotron facility is required for both dispersive-NEXAFS and rapid-XPS. Indeed, there are few alternatives to SDR and RDS for real-time measurement with high sensitivity in the laboratory, and they also offer sufficiently to investigate surface chemical reaction processes.

A key advantage of real-time measurements is that they allow flexible adjustment of reactions by changing the rate of the reaction or stopping it based on in situ monitoring of the growth process. In order to understand surface reaction processes, it is useful to analyze the temperature dependence of the reaction time to estimate the activation energies. For example, Yasuda et al. have shown that it is possible to detect layer-by-layer oxidation on Si(001) as an RD oscillation, and, they also evaluated the activation energies for the growth of two to four oxide layers using the RD oscillation method [11]. In this article, we introduce our recent studies on the initial oxidation process of Si(001) in the monolayer regime by means of both SDR [12–14] and RDS [15]. Here, monolayer oxidation on Si(001) means that the surface is oxidized up to the backbond of the dimer.

2.2 Experimental

Experiments were performed with an ultrahigh vacuum chamber, whose base pressure was under 2.0×10^{-8} Pa. The sample was a p-type Si(001) single crystal with resistivity of 12–$14\,\Omega$cm, cut to the size of $3 \times 20 \times 0.63\,\mathrm{mm}^3$. We used a Si(001) wafer with a miscut at $4°$ toward the [110] direction to acquire single-domain Si(001)-(2×1) structure. Anisotropy of the surface structure is required to obtain the RD signal. The sample was well degassed for over 12 h at 873 K and cleaned by flashing at 1470 K for 10 s. It was then subjected to oxidation with molecular O_2 at a pressure of 1.0×10^{-5} Pa in the temperature range of 300–923 K for SDR and 583–923 K for RDS.

2.2.1 SDR Measurement

SDR using p-polarized light at the **Brewster angle** was developed by Kobayashi et al. around 1990 [2]. In SDR, the intensity of the difference spectra is defined as

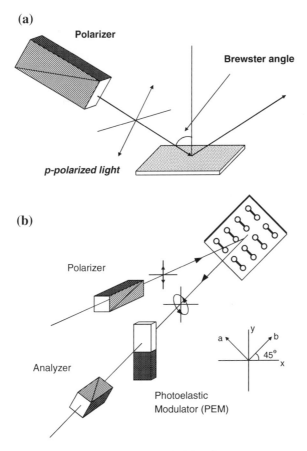

Fig. 2.1 a Schematic of SDR optics. **b** Schematic of RDS optics

$$\frac{\Delta R}{R} = \frac{R_a - R_c}{R_c} \tag{2.1}$$

where R_c and R_a represent the reflectance of the clean surface and that of the oxygen-adsorbed surface, respectively. The intensity of the reflected light from the bulk is minimum at the Brewster angle for incident p-polarized light [16]. Hence, the SDR signal corresponding to the structural changes at the surface can be obtained by using p-polarized light at the Brewster angle. This method was proved to be more surface sensitive than an earlier version known as **differential reflectance spectroscopy (DRS)**, which employed normal incidence.

A schematic of the SDR optics is shown in Fig. 2.1a. Figure 2.2 shows a schematic of the SDR-RDS apparatus constructed for this work. A halogen lamp (1.3–2.6 eV) was used as the light source. To eliminate second-order light, the light reflected through a grating in the monochromator was filtered above 2.6 eV. SDR measurement

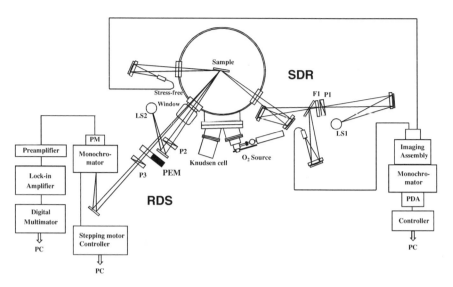

Fig. 2.2 Experimental setup for SDR and RDS

does not have any requirement for surface anisotropy, in contrast to RDS. We used a **photodiode array** in order to detect the whole spectrum at the same time during real-time measurement. The SDR signal was averaged over 10 s in this experiment.

The advantage of SDR over RDS is the ability to detect the whole spectrum instantaneously. It is sometimes difficult to measure the signal when changes of the surface structure or electronic states are small. For example, it is important to identify any spurious drift component, especially at high temperatures.

2.2.2 RDS Measurement

The optical setup for the RD measurements was designed following the configuration reported by Aspenes et al. [1, 17]. A schematic of the RDS optics is shown in Fig. 2.1b. We used a **strain-free window** to minimize optical anisotropy irrelevant to the surface structure. A **xenon lamp** was used as the light source from the visible to near-ultraviolet region (2.0–5.0 eV).

The RD amplitude, $\Delta\tilde{r}/\tilde{r}$ is defined as

$$\frac{\Delta\tilde{r}}{\tilde{r}} = \frac{(\tilde{r}_a - \tilde{r}_b)}{(\tilde{r}_a + \tilde{r}_b)/2} \tag{2.2}$$

where $\tilde{r}_a$ and $\tilde{r}_b$ are complex reflectance for polarization parallel (a-axis) and perpendicular (b-axis) to the direction of the dimer bonds on a clean Si(001)-(2×1) surface, respectively. The RD amplitude can be rewritten as

Fig. 2.3 Schematic of the
three-phase model

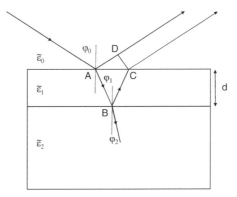

$$\frac{\Delta \tilde{r}}{\tilde{r}} = \frac{\Delta r}{r} + i \cdot \Delta \theta \tag{2.3}$$

In RDS measurement, the surface must exhibit anisotropy in order to obtain a signal. It is known that the accuracy of measurement is higher for the real part ($\Delta r/r$) than the imaginary part ($\Delta \theta$) [4]. The **Kramers-Kronig relation** holds between these quantities. Hence, substantial information can be extracted from the analysis of the real part. We present here the experimental results for the real part.

In the case of RDS, the time course of the RD intensity is measured at a fixed wavelength (energy) in the usual optical setup. Hence, only information at a certain wavelength can be obtained by a single measurement. The RD signals are measured ten times and average values are plotted in the present work.

It takes several minutes to obtain the RD spectrum, and this is a limitation of the method together with the requirement for anisotropy of the surface structure. The limitation of the fixed wavelength can be surmounted by means of multichannel simultaneous measurements (rapid RDS) [18]. Further details and applications of SDR and RDS have been covered in review articles [3, 4].

2.2.3 Formulation of the Reflectance in a Three-Phase Model

A principles of RDS and SDR have been well described in textbooks [19–21]. The general approach for analyzing the reflectance spectra is common to spectroscopic ellipsometry (SE) [22]. Here, we briefly introduce the three-phase model.

Figure 2.3 shows the three phase model, which consists of vacuum (0), thin film (1), and substrate (2). These numbers are used as subscripts of the parameters in the following. Important parameters are complex **dielectric constant** $\tilde{\epsilon}_0$, $\tilde{\epsilon}_1$, $\tilde{\epsilon}_2$ and the film thickness d. Magnetic permeabilities are $\mu_0 = \mu_1 = \mu_2 = 1$ in the optical frequency range. We define $\tilde{r}_{ij}$ as the **Fresnel coefficient** of the interface between phase i and j, and $\tilde{r}_{ijk}$ as representing the Fresnel coefficient for the reflected light into the vacuum, including multiple reflection:

$$\tilde{r}_{012}(\tilde{\beta}) = \frac{\tilde{r}_{01} + \tilde{r}_{12}e^{-2i\tilde{\beta}}}{1 + \tilde{r}_{01}\tilde{r}_{12}e^{-2i\tilde{\beta}}} \qquad (2.4)$$

The film phase thickness $\tilde{\beta}$ is given by

$$\tilde{\beta} = \frac{2\pi d\tilde{n}_0 \cos\varphi_0}{\lambda} \qquad (2.5)$$

where the complex refractive index is $\tilde{n}_0 = \sqrt{\epsilon_0} = 1$. In the analysis of a system such as the solid/liquid interface, the complex refractive index should be used. The normalized reflectivity change in SDR [23] is given by

$$\frac{\Delta R}{R} = \frac{R_{012}(\tilde{\beta}) - R_{012}(\tilde{\beta} = 0)}{R_{012}(\tilde{\beta} = 0)} \qquad (2.6)$$

$$= \frac{|\tilde{r}_{012}(\tilde{\beta})|^2 - |\tilde{r}_{012}(\tilde{\beta} = 0)|^2}{|\tilde{r}_{012}(\tilde{\beta} = 0)|^2} \qquad (2.7)$$

Fresnel coefficient $\tilde{r}_{ij}$ is given by the **Fresnel formula**, which is different for s-polarized light and p-polarized light. If the thickness is very much less than the wavelength $(d/\lambda \ll 1)$, the Fresnel coefficient of the three-phase system can be expanded to terms of first order in $\tilde{\beta}$

$$\frac{\Delta R_s}{R_s} = \frac{8\pi d \cos\varphi_0}{\lambda} \left(\frac{\tilde{\epsilon}_1 - \tilde{\epsilon}_2}{1 - \tilde{\epsilon}_2} \right) \qquad (2.8)$$

$$\frac{\Delta R_p}{R_p} = \frac{8\pi d \cos\varphi_0}{\lambda} \left[\left(\frac{\tilde{\epsilon}_1 - \tilde{\epsilon}_2}{1 - \tilde{\epsilon}_2} \frac{1 - (1/\tilde{\epsilon}_1\tilde{\epsilon}_2)(\tilde{\epsilon}_1 + \tilde{\epsilon}_2)\sin^2\varphi_0}{1 - (1/\tilde{\epsilon}_2)(1 + \tilde{\epsilon}_2)\sin^2\varphi_0} \right) \right] \qquad (2.9)$$

In the case of RDS, normal incidence $(\varphi_0 = 0)$ is applied and the Fresnel coefficient of the three-phase system [24] is given by

$$\tilde{r}_{012}(\tilde{\beta}; \tilde{\epsilon}_1) = \tilde{r}_{012}(\tilde{\beta} = 0)\left(1 - 2i\tilde{\beta}\frac{\tilde{\epsilon}_1 - \tilde{\epsilon}_2}{\tilde{\epsilon}_0 - \tilde{\epsilon}_2} \right) \qquad (2.10)$$

Complex reflectances along the a-axis and the b-axis can be approximated by inserting the diagonal elements $(\tilde{\epsilon}_a, \tilde{\epsilon}_b)$ of the complex dielectric constant. By taking the first order term in $\tilde{\beta}$

$$\frac{\Delta \tilde{r}}{\tilde{r}} = \frac{\tilde{r}_{012}(\tilde{\beta}; \tilde{\epsilon}_a) - \tilde{r}_{012}(\tilde{\beta}; \tilde{\epsilon}_b)}{(\tilde{r}_{012}(\tilde{\beta}; \tilde{\epsilon}_a) + \tilde{r}_{012}(\tilde{\beta}; \tilde{\epsilon}_b))/2} \qquad (2.11)$$

$$= -\frac{4\pi\tilde{n}_0 d}{\lambda} \cdot \frac{\tilde{\epsilon}_a - \tilde{\epsilon}_b}{\tilde{\epsilon}_0 - \tilde{\epsilon}_2} \qquad (2.12)$$

Fig. 2.4 Schematic of representative adsorption sites on Si(001)-(2×1)

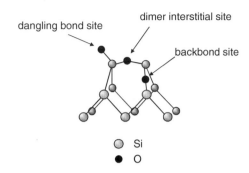

This formula can be applied to a solid/liquid interface with complex refractive index $\tilde{n}_0$ and complex dielectric constant $\tilde{\epsilon}_0$. The above equations representing the differential reflectance signal of SDR and RDS are called the **McIntyre-Aspnes formula**.

In the formulation based on the three-phase model, nonlocality of the complex dielectric tensor $\tilde{\epsilon}_{ij}(r, r'; \omega)$, structure of the surface/interface, and inhomogeneity of the film were ignored. There are more sophisticated approaches [25–27] connecting reflectance to complex dielectric tensor. However, the McIntyer-Aspnes formula can be derived from them by appropriate approximation. Therefore, the McIntyer-Aspnes formula is still routinely used at present.

2.2.4 Advantages of the Use of SDR and RDS

Figure 2.4 shows typical adsorption sites on Si(001) in the monolayer regime. Our results with SDR indicate that partial oxygen coverages at different adsorption sites can be evaluated [12–14]. Partial oxygen coverage at a specific adsorption site is an important parameter for evaluation of the mechanism of oxidation processes. This is because the stoichiometry of the oxide film, the strain induced at the interface between the oxide film and the silicon substrate and work function are directly associated with partial oxygen coverage [28, 29]. In the analysis of the oxide formation on Si(001), the Si 2p core-level oxide components (Si^+, Si^{2+}, Si^{3+}, Si^{4+}) have been intensively studied [30]. Such analyses provide information on the density of silicon atoms attached to a certain number of oxygen atoms. However, this does not mean that partial oxygen coverages can be determined straightforwardly. In addition, it is difficult to directly observe an oxygen atom adsorbed at the back-bond site by means of scanning tunneling microscopy (STM) [31]. Therefore, estimation of partial oxygen coverages by means of SDR and RDS will provide us with complementary information on the oxidation mechanism.

Tanaka et al. evaluated activation energies for adsorption and desorption of halogen atoms on Si(111)-(7×7) by SDR and also determined the order of the reaction process [6]. As already mentioned, Yasuda et al. evaluated activation energies for the

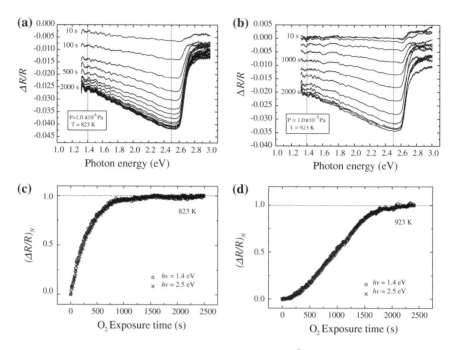

Fig. 2.5 SDR spectra during oxidation at a pressure of 1.0×10^{-5} Pa at the temperature of **a** 823 K and **b** 923 K. Time courses of SDR intensity at **c** 823 K and **d** 923 K. These uptake curves are normalized so that the intensity at the beginning and the end point are zero and one, respectively

growth of two to four oxide layers on Si(001) by RDS [11]. We applied both techniques to examine activation energies at specific adsorption sites in the monolayer oxidation process.

2.3 Analysis of Initial Oxidation Reaction on Si(001)

2.3.1 SDR Results

Figure 2.5 shows the time course of the SDR intensity observed for oxidation at 823 and 923 K. The uptake curves at 1.4 and 2.5 eV are normalized so that the intensities at the beginning and the end point are zero and one, respectively. We measured spectra at the interval of 10 s. In the figures, we present spectra at the interval of 100 or 200 s for clarity. No distinct peaks were observed at the photon energy of 1.3–2.6 eV. The normalized time courses show no photon energy dependence at both 823 and 923 K. However, the line shape of the time courses is markedly different between these temperatures. The exponential curve at 823 K represents Langmuir-type adsorption,

Fig. 2.6 Time courses of SDR intensity at the temperature of 300 K, taken at the photon energies of 1.4 and 2.5 eV. The uptake curve is normalized so that the intensity at the beginning and the end point are zero and one, respectively

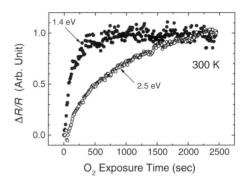

while the sigmoid curve at 923 K corresponds to two-dimensional island growth. We investigated two photon energies (1.4 and 15 eV) since distinct photon energy dependence was observed at room temperature (RT).

The change of the line shape is consistent with the results obtained by **ultraviolet photoemission spectroscopy (UPS)** [32] and **Auger electron spectroscopy (AES)** [33], indicating that the SDR intensity is almost proportional to the total oxygen coverage. The time course for the Langmuir-type adsorption can be fitted with a single-exponential function:

$$I(t) = 1 - \exp(-\kappa_1 t) \qquad (2.13)$$

where κ_1 represents the reaction rate. The time course should be normalized before fitting.

The photon energy dependence at RT (300 K) is shown in Fig. 2.6. It is apparent that the line shapes are different between 1.4 and 2.5 eV. The 1.4 eV component increases rapidly while the 2.5 eV component increases slowly.

It is known that the back-bond site is the most stable when one oxygen molecule is dissociatively adsorbed on the silicon dimer of Si(001)-(2 × 1) [34]. Hence, a possible assignment is that the features at 1.4 and 2.5 eV correspond to the back-bond site and dangling-bond site or dimer-interstitial site of the silicon dimer, respectively. At present, there are no theoretical calculations supporting this assignment.

However, it seems reasonable to consider that the reaction rate should be faster when the final adsorption site is more stable. Nevertheless, we have to consider various intermediate structures and activation barriers before the adsorbate reaches the stable adsorption site. We therefore use tentative notation, designating 1.4 and 2.5 eV components as α-state and β-state, respectively. From the analysis of the temperature dependence of κ_1, the activation energies are estimated to be $\epsilon_\alpha = 0.04 \pm 0.02$ and $\epsilon_\beta = 0.11 \pm 0.02$ eV. The initial slope ($I(t) \leq 0.3$) was examined to evaluate the activation energies at the initial stage of oxidation. The estimated values are $\epsilon_\alpha^* = 0.04 \pm 0.02$ eV and $\epsilon_\beta^* = 0.12 \pm 0.02$ eV.

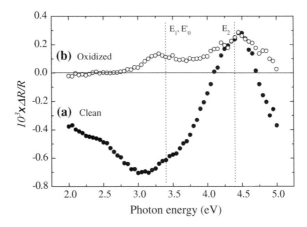

Fig. 2.7 a RD spectrum of a clean Si(001)-(2 × 1) surface with 4° miscut toward the [110] direction. **b** RD spectrum of the surface oxidized at the oxygen pressure of 1.0×10^{-5} Pa and at the temperature of 583 K

2.3.2 RDS Results

Figure 2.7 shows typical RD spectra obtained for a clean Si(001)-(2×1) surface and an oxidized surface. Here, the surface was oxidized at the pressure of 1.0×10^{-5} Pa and at the temperature of 583 K. A shoulder-like structure at 3.4 eV, corresponding to the E_1 and E_0' critical points, is the fingerprint of a single-domain Si(001)-(2×1) surface with a 4° miscut angle [35]. In Fig. 2.8, we present the time course of RD intensity for oxidation at 583 and 873 K taken at the photon energy of 3.1 eV. The oxygen uptake curves with SDR exhibit different line shapes, i.e., a sigmoid curve (two-dimensional island growth) and a single exponential curve (Langmuir-type adsorption). On the other hand, the time courses observed with RD measurements could be fitted with a single exponential curve at both temperatures. The increase rate of the total oxygen coverage is small at the beginning of two-dimensional island growth, while that of the RD intensity is very large.

Typical features on the oxidized surface in the monolayer regime are the positive peaks at E_1, E_0' (3.4 eV), and E_2 (4.4 eV). Fuchs et al. showed that oxygen inserted into Si-Si bonds in the ($\bar{1}$10) or (110) plane reproduces the RD features of the oxidized surface [36]. Moreover, they showed that uniformly compressed deformation of the Si lattice by 2 % along [$\bar{1}$10] could roughly reproduce the RD features of the oxidized surface. Therefore, the RD features of the oxidized surface can be ascribed to deformation of the Si lattice.

The relationship between the RD intensity and the oxygen coverage is not so clear. We showed that all the time courses at 583~823 K could be well fitted with a single exponential function. It should be noted that SiO desorption does not occur in the Langmuir-type adsorption regime at temperatures below 823 K.

Based on our present results, it can be assumed that the change of RD intensity is almost proportional to the total oxygen coverage in the Langmuir-type adsorption

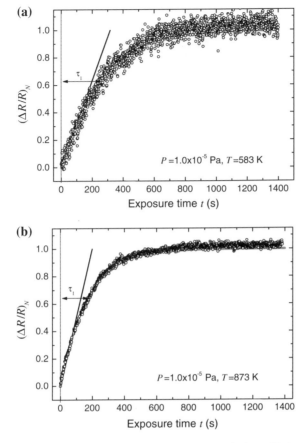

Fig. 2.8 Time courses of RD intensity at 3.1 eV measured in situ during oxidation at a pressure of 1.0×10^{-5} Pa. Lines represent linear fitting of the plots for $\Delta R/R \leq 0.3$. The growth modes are **a** Langmuir-type adsorption at 583 K and **b** two-dimensional island growth at 873 K

regime. In the case of two-dimensional island growth, perturbation of the Si lattice must be larger because the Si atoms on the surface are partly expelled in the SiO desorption process. We obtained the activation energy $\epsilon_1 = 0.16 \pm 0.03$ eV in the temperature range of 583 ~ 823 K. From analysis of the initial slope, the activation energy at low coverage is estimated to be $\epsilon_1^* = 0.26 \pm 0.03$ eV.

2.3.3 Comparison of the Results Obtained with SDR and RDS

In the analysis of the activation energies with SDR, the value obtained by the fitting of the whole uptake curve and that obtained by the fitting of the initial slope were the same within the limits of error. On the other hand, the activation energy obtained by the fitting of the initial slope is larger than that obtained by the fitting of the whole uptake curve in the case of RDS. It is known that Si dimer ejection occurs during thermal oxidation [31]. It is possible that the defect produced by Si dimer ejection acts as an active site for further oxidation, giving rise to coverage dependence of the activation energy.

However, it is not easy to interpret the different features in the coverage dependence obtained with SDR and RDS. The reason may be related to the difference in the quantities observed with these techniques: SDR is sensitive to changes of the local bonding structure, while RDS detects anisotropy predominantly induced at the SiO_2/Si interface. It may be possible to construct a reaction model which can explain the apparent discrepancy between the results obtained with SDR and RDS, by taking account the diffusion of Si atoms at high temperature.

2.3.4 Origin of the Activation Energy

In this study, investigation of the existence of activation energy in the mono-layer regime is the main focus of attention. Assuming the existence of a finite activation energy, evaluation of the activation energies at different adsorption sites and at different oxygen coverages becomes an important problem.

It has been theoretically proposed that the initial interaction of an oxygen molecule at the buckled dimer of Si(001) predominantly mediates adsorption of oxygen at the back-bond sites through barrierless dissociation [34]. A recent theoretical study revealed that an activation energy of $0.05 \sim 0.11$ eV may exist when the incoming molecule does not take the geometry parallel to the dimer [37]. The activation barrier at the crossing point of the potential energy surfaces of the triplet state, which is the ground state for an isolated oxygen molecule, and the singlet state was considered, and the estimated activation energies were similar to the values obtained in our experimental work.

An early UPS study showed that O(2p)-derived valence band intensity curves exhibit no temperature dependence at $623 \sim 827$ K [38]. According to an AES study combined with **scanning reflectance electron microscopy (SREM)**, AES intensity curves showed no significant change at RT-473 K [39]. These results support a barrierless dissociation model. On the other hand, recent STM investigation on a stepped Si(001)-c(4×2) surface at 80 K suggested that mobile oxygen species with a long lifetime should exist on the surface and migrate until they are adsorbed near the step edges. In this case, a finite activation energy for dissociative adsorption on a flat terrace should exist, at least at low temperature.

Here, we divide the problem into two parts. First, we can pose two questions: (1) Is there a finite activation energy when a single oxygen molecule is interacted with a clean Si(001) surface? (2) What is the molecular precursor state which can migrate on the surface before dissociative adsorption? Second, irrespective of the existence of a finite activation energy at low coverage limit, the value of the activation energy may change with the oxygen coverage.

The results of SDR, RDS, and STM studies support a mechanism with a finite activation barrier at the low coverage limit. However, we have to be cautious, because these results are inconsistent with previous results obtained with UPS and AES. It is possible that the early studies overlooked the slight temperature dependence associated with a finite activation energy of about 0.1 eV. We also have to consider that each experimental method may give information on different aspects of the oxidation process.

The estimated activation values $0.04 \sim 0.26$ eV are less than the activation energies $\epsilon_2 = 1.2$ eV [11] ($\epsilon_2^* = 0.3$ eV [39]) for the growth of the second layer. Here, the asterisk represents the value at the initial stage, in the monolayer regime. It is known that the growth rate in the first layer is much faster than that in the second layer, indicating a smaller value of the activation energy in the monolayer regime.

In theoretical calculations, interaction of one oxygen molecule at a clean Si(001) surface has often been investigated. Therefore, correspondence with the experimental findings may not be straightforward in most cases. Not only the experimentally evaluated activation energies with SDR and RDS, but also those with UPS and AES represent the *average* value for the growth of each layer.

Modeling of the SiO_x/Si interface on a large scale has been attempted by means of first-principles molecular dynamics [40]. Such a method, involving many atoms, should be suitable to elucidate the coverage dependence of the activation energy in detail. Investigation of the reaction with a single molecule at low coverage will require state-of-the-art experimental methods such as molecular beam techniques.

To summarize, the present results with SDR and RDS indicate that a small but finite activation energy exists in monolayer oxidation, not only on average over the whole process, but also at low coverage.

2.3.5 Qualitative Aspects of the Reflectance Spectra

In the qualitative analysis of reflectance spectra, a spectral feature due to the surface state or the electronic state of an adsorbate is assigned to be *extrinsic*, while that due to perturbation of the bulk state by adsorption is assigned to be *intrinsic*. In the case of Si, the effect of the interband transition appears at the energies above 3.2 eV [41].

According to first-principles calculations [36], the surface state completely disappears on a Si(001) surface covered with a monolayer oxide. Hence, the reflectance spectra have their intrinsic origin at the oxygen-covered surface. Here, the underlying physical picture can be described as follows. (1) The silicon lattice is deformed by oxygen adsorption. (2) The electronic state of the silicon substrate is modified predominantly by the strain effect. (3) As a result, the intrinsic dielectric property changes, resulting in changes of the reflectance signal. In the present work, the energy dependence observed in SDR at energies below 2.6 eV can be ascribed to changes of the surface state, that is, its origin is extrinsic. Discrimination between the surface state and the bulk state as the origin of a spectral feature is often not straightforward. In the case of RDS, Nakayama et al. described the qualitative features of the reflectance spectra on silicon surfaces in detail [42].

2.3.6 Adsorption Structure, Identification of the Adsorption Sites, and Characterization of the Interface Structure

It is difficult to evaluate partial coverages of oxygen adsorbed on a Si(001) surface in the monolayer regime. As already mentioned, it is difficult to identify oxygen adsorbed at the back-bond site by means of STM [31]. In the present work, we show that the partial coverages can be evaluated by analyzing the energy dependence of the SDR spectra.

It is reported that adsorption on the adatom and breaking of the bond could be identified separately and simultaneously in hydrogenation [5] and adsorption/desorption of halogen species [6] by means of SDR. Following these studies, Borenstein et al. showed that spectral lineshape at 3.5 eV in the SDR spectra on Si(001) is sensitive to adsorption or bond breaking at the dimer bond [7]. We showed that reflectance spectra on chlorine-adsorbed Si(111) can be reproduced by cluster model calculations [43]. At the energy range below 3.2 eV, the cluster model with a small number of atoms is applicable because the bulk state is not important.

Analysis using spectral decomposition can be applied to the RD spectra at RT where no SiO desorption occurs, in order to evaluate partial coverages of specific oxide structures [44]. The results of this study support preferential adsorption at the back-bond site, in agreement with the present work (Fig. 2.6). In the analysis of the SiO_2/Si interface with a thick oxide layer, the RD signal from the SiO_2 film with amorphous structure or from the bulk is reduced, which enables us to obtain information about the interface, in principle. In this sense, RDS is suitable to observe reaction processes occuring at the buried interface effectively. We recently evaluated the morphology of the SiO_2/Si interface on high-index silicon surfaces by means of RDS [45].

Theoretical calculations are indispensable to analyze local adsorption sites or local structures based on the reflectance spectra obtained with SDR and RDS. It is desirable to compare the results with those obtained using other surface analysis techniques, as described in the present work. It has been well established that SDR and RDS show sufficient surface sensitivity to observe reaction processes in the submonolayer regime. Most of the calculations of the reflectance spectra deal with relatively simple systems. Recently, however, attempts have been made to describe an organic-molecule adsorbed system [46]. Extension of both experimental and theoretical work in parallel will further extend the applicability of these optical methods.

2.4 Summary and Outlook

In this article, we have introduced our recent studies utilizing two linear optical methods, that is, surface differential reflectance (SDR) spectroscopy and reflectance difference spectroscopy (RDS). We focused on application of these methods to the initial oxidation process on Si(001) in detail. Identification of the growth modes (Langmuir-type adsorption and two-dimensional island growth), evaluation of the transition temperature, and evaluation of activation energies were described.

SDR and RDS are versatile techniques which can be applied under various conditions from ultrahigh vacuum to atmospheric, as well as to solid–liquid interfaces as an experimental tool for in situ and *real-time* measurements. Optical methods is not only linear optical spectroscopy, but also nonlinear optical spectroscopy such as second harmonic generation (SHG) are powerful *non-destructive* measurement tools to investigate epitaxial growth processes in combination with other surface/interface analysis tools. The term "epioptics" has been used to represent this growing field of research.

In the field of renewable energy research, there is considerable interest in organic thin films. Real-time measurements of the growth process of organic thin films are now being conducted by our research group, and we hope this will extend further the applicability of both SDR and RDS.

2.5 Acknowledgement

We would like to thank Dr. T. Yasuda of the National Institute of Advanced Industrial Science and Technology (AIST) for valuable help in constructing the RDS apparatus. We would like to thank Prof. T. Suzuki of the National Defence Academy for his support at the initial stage of the present work.

References

1. D.E. Aspnes, J.P. Harbison, A.A. Studna, L.T. Florez, J. Vac. Sci. Technol. A **6**, 1327 (1988)
2. N. Kobayashi, Y. Horikoshi, Jpn. J. Appl. Phys. **28**, L1880 (1989)
3. J.F. McGilp, Prog. Surf. Sci. **49**, 1 (1995)
4. P. Weightman, D.S. Marin, R.J. Cole, T. Farrell, Rep. Prog. Phys. **68**, 1251 (2005)
5. C. Beitia, W. Preyss, R. Del Sole, Y. Borensztein, Phys. Rev. B **56**, R4371 (1997)
6. M. Tanaka, E. Yamakawa, T. Shirao, K. Shudo, Phys. Rev. B **68**, 165411 (2003)
7. Y. Borensztein, O. Pluchery, N. Witkowski, Phys. Rev. Lett. **95**, 117402 (2005)
8. J. Kondoh, I. Nakai, M. Nagasaka, K. Amemiya, T. Ohta, J. Vac. Soc. Jpn. **52**, 73 (2009)
9. M. Okada, Y. Teraoka, J. Vac. Soc. Jpn. **52**, 80 (2009)
10. A.R. Turner, M.E. Pemble, J.M. Fernandez, B.A. Joyce, J. Zhang, A.G. Taylor, Phys. Rev. Lett. **74**, 3213 (1995)
11. T. Yasuda, N. Kumagai, M. Nishizawa, S. Yamasaki, H. Oheda, K. Yamabe, Phys. Rev. B **67**, 195338 (2003)

12. J. Takizawa, J. Koizumi, S. Ohno, K. Shudo, M. Tanaka, Shinku **49**, 323 (2006)
13. J. Takizawa, S. Ohno, J. Koizumi, K. Shudo, M. Tanaka, J. Phys.: Condens. Matter **18**, L209 (2006)
14. S. Ohno, J. Takizawa, J. Koizumi, F. Mitobe, R. Tamegai, T. Suzuki, K. Shudo, M. Tanaka, J. Phys.: Condens. Matter **19**, 446011 (2007)
15. S. Ohno, H. Kobayashi, F. Mitobe, T. Suzuki, K. Shudo, M. Tanaka, Phys. Rev. B **77**, 085319 (2008)
16. Y. Horikoshi, M. Kawashima, N. Kobayashi, J. Cryst. Growth **111**, 200 (1991)
17. T. Yasuda, Kotaibutsuri **34**, 99 (1999) (in Japanese)
18. P. Harrison, T. Farrell, A. Maunder, C.I. Smith, P. Weightman, Meas. Sci. Technol. **12**, 2185 (2001)
19. H. Lüth, *Solid Surfaces, Interfaces and Thin Films* (Springer, Berlin, 2001)
20. V.G. Bordo, H.-G. Rubahn, *Optics and Spectroscopy at Surfaces and Interfaces* (WILEY-VCH Verlag GmbH & Co. KGaA, Weinheim, 2005)
·21. T. Onishi, Y. Horiike, K. Yoshihara (eds.), *Kotaihyomenbunseki II* (Kodansha Scientific, 1995) (in Japanese)
22. H. Fujiwara, *Spectroscopic Ellipsometry* (Maruzen 2003) (in Japanese)
23. J.D.E. McIntyer, D.E. Aspnes, Surf. Sci. **24**, 417 (1971)
24. K. Hingerl, D.E. Aspnes, I. Kamiya, L. Florez, Appl. Phys. Lett. **63**, 885 (1993)
25. M. Nakayama, J. Phys. Soc. Jpn. **39**, 265 (1975)
26. A. Bagchi, R.G. Barrera, A.K. Rajagopal, Phys. Rev. B **20**, 4824 (1979)
27. R. Del Sole, Solid State Commun. **37**, 537 (1981)
28. S. Ogawa, Y. Takakuwa, Jpn. J. Appl. Phys. **44**, L1048 (2005)
29. S. Ogawa, A. Yoshigoe, S. Ishidzuka, Y. Teraoka, Y. Takakuwa, Jpn. J. Appl. Phys. **46**, 3244 (2007)
30. F.J. Himpsel, F.R. McFeely, T. Taleb-Ibrahimi, J.A. Yarmoff, G. Hollinger, Phys. Rev. B **38**, 6084 (1988)
31. Ph Avouris, C. Cahill, Ultramicroscopy **42–44**, 838 (1992)
32. M. Suemitsu, Y. Enta, Y. Miyanishi, N. Miyamoto, Phys. Rev. Lett. **82**, 2334 (1999)
33. Y. Takakuwa, F. Ishida, T. Kawawa, Appl. Surf. Sci. **190** (2002) 20; **216** (2003) 133
34. K. Kato, T. Uda, K. Terakura, Phys. Rev. Lett. **80**, 2000 (1998)
35. N. Witkowski, R. Coustel, O. Pluchery, Y. Borensztein, Surf. Sci. **600**, 5142 (2006)
36. F. Fuchs, W.G. Schmidt, F. Bechstedt, Phys. Rev. B **72**, 075353 (2005)
37. X.L. Fan, Y.F. Zhang, W.M. Lau, Z.F. Liu, Phys. Rev. Lett. **94**, 016101 (2005)
38. Y. Enta, Y. Takegawa, M. Suemitsu, N. Miyamoto, Appl. Surf. Sci. **100**(101), 449 (1996)
39. H. Watanabe, K. Kato, T. Uda, K. Fujita, M. Ichikawa, T. Kawamura, K. Terakura, Phys. Rev. Lett. **80**, 345 (1998)
40. A. Pasquarello, M. Hybertsen, R. Car, Nature **396**, 58 (1998)
41. A. Cricenti, J. Phys.:Condens. Matter **16** (2004) S4243
42. T. Nakayama, M. Murayama, T. Yasuda, Kotaibutsuri **38**, 201 (2003) (in Japanese)
43. Y. Mogawa, S. Ohno, K. Shudo, M. Tanaka, J. Vac. Soc. Jpn. **49**, 138 (2006) (in Japanese)
44. K. Gaal-Nagy, A. Incze, G. Onida, Y. Borensztein, N. Witkowski, O. Pluchery, F. Fuchs, F. Bechstedt, R. Del Sole, Phys. Rev. B **79**, 045312 (2009)
45. S. Ogata, S. Ohno, M. Tanaka, T. Horikawa, T. Yasuda, Jpn. J. Appl. Phys. **49**, 022403 (2010)
46. K. Seino, W.G. Schmidt, Surf. Sci. **548**, 183 (2004)

Chapter 3
Development of In-Vacuum Microscope Under High Pressure and Its Applications

Takao Nanba

3.1 Introduction

Solids under pressure suffer from a drastic change in physical and chemical properties according to temperature and pressure. Photoemission and tunneling spectroscopies which have been recognized as a very powerful tool to study an electronic state close to a Fermi level of solids do not work under pressure. Then optical measurement seems to be the most effective method among various kinds of experiments to investigate electronic states of solids under pressure. Recently, new materials which exhibit attractive physical and chemical properties under pressure have been successfully synthesized according to an advance in experimental technology in various fields of science. However, most of synthesized samples are not always single crystals but polycrystals. In such cases, new compounds suffer from inhomogeneous distribution of constituent elements and then we need a local analysis on small-area sample surface by a two-dimensional (2D) imaging spectroscopy in order to investigate electronic states. Most suitable experiment to answer such requirements is spectromicroscopy on solids under pressure.

At present, spectromicroscopic experiments are available in wavelength regions from ultraviolet to infrared light, but not in far-infrared/terahertz (THz) wave because of experimental difficulties. In general, a generation of pressure in GPa range in spectroscopic experiments on solids under pressure is performed by using a diamond anvil cell (DAC) which is shown in Fig. 3.1a. The transmission spectrum of a type IIa of diamond thin film in the regions from a near-infrared to an infrared light is shown in Fig. 3.1b. The absorption edge of diamond is almost 300 nm and then a DAC is almost transparent in the whole regions from ultraviolet to far-infrared/THz wave except around $2000\,cm^{-1}$. Consequently, spectromicroscopic experiments combined with a DAC is available in the regions from ultraviolet to far-infrared/THz wave.

T. Nanba (✉)
Kobe University, 4-52-201 Nishiyama-cho, Kohyohen, Nishinomiya 662-0017, Japan
e-mail: hfg78301@hcc6.bai.ne.jp

K. Shudo et al. (eds.), *Frontiers in Optical Methods*,
Springer Series in Optical Sciences 180, DOI: 10.1007/978-3-642-40594-5_3,
© Springer-Verlag Berlin Heidelberg 2014

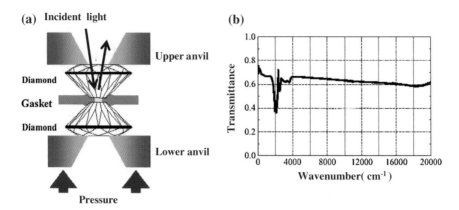

Fig. 3.1 **a** A cross sectional view of a diamond anvil cell (*DAC*) indicating an optical configuration of reflection measurement. **b** Transmission spectrum of a IIa type of diamond thin film. Sharp dip structure around $2000\,\mathrm{cm}^{-1}$ is due to two phonon absorption in diamond

Particularly, it must be a very powerful tool in infrared and far-infrared/THz wave light to investigate a change in an electronic state close to a Fermi level of solids under high pressure because the photon energies of an infrared to a far infrared/THz wave lights are comparable with a scale of a fine electronic structure around a Fermi level of newly synthesized compounds.

In this article, we focus our interest on spectromicroscopy combined with a DAC as a method to study optical properties of solids under pressure. We developed a new type of microscope, named as "In-ultra-high-vacuum (in-UHV) microscope" [1]. The name of the In-UHV microscope means that all optical components of microscope are installed in the ultra high vacuum (UHV) chamber and separated by CVD diamond windows from the low-vacuum parts of the optical system including of the spectrometer. The developed microscope covers very wide wavelength regions, not only from infrared to far-infrared/terahertz wave but also from visible to ultraviolet light, and make possible the spectromicroscopic measurements in the whole spectral region without any replacement of the optical windows of the microscope. First, we describe the details of the fundamental concepts and the design of the in-UHV microscope, and second introduce the results of the studies on ice, $Nd_2Ir_2O_7$ and $Pb_2Ir_2O_7$ under high pressure at low temperature. Ice under pressure undergoes an order-disorder phase transition on the hydrogen bond and has attracted considerable interest as the prototype of hydrogen-bonded materials. And $Nd_2Ir_2O_7$ is one of the representative compound of Ir-based pyrochlore oxides $Ln_2Ir_2O_7$ (Ln = rare earth elements) which have recently attracted very wide research interest [2] because they exhibit very attractive phenomena as mentioned below. $Pb_2Ir_2O_7$ is the compound which was recently successfully synthesized and considered to be the most important reference material to investigate the role of $Ln - 4f$ electrons of Ln-ion in the physical properties of $Ln_2Ir_2O_7$.

3.2 Development of In-UHV Microscope

3.2.1 Why In-UHV Microscope

Almost of conventional microscopes work only in atmospheric environments or at least they are a purged type which means that the optical pass of the microscope is filled with dry nitrogen gases. However, such types of microscopes seem to be very inconvenient as a tool to investigate optical properties of solids under pressure at low temperature by following two reasons. (1) Spectroscopy on solids under pressure at low temperature requires the specromicroscopic system combined with a low temperature cryostat which is equipped inevitably with different kinds of optical windows in each available wavelength region. This means that a frequent replacement of optical windows of microscopic system is required according to different wavelength regions. More than all, optical study on materials which exhibit metallic properties requires a reflection (R) measurement in a very wide wavelength region because optical constants of metallic materials are obtained from a Kramers–Kronig (K–K) transformation process of measured R-spectra and in principle a K–K transformation needs the R-spectrum in the energy region of [0, ∞] [3]. However, the R-measurement by conventional microscopes is limited in the regions from visible to infrared light because of the experimental limitations. Then, in an actual K–K process, we measure first the spectra in the wide range from visible to infrared light and secondly perform the K–K analysis by using of suitable extrapolation functions at the both parts of low and high energy regions of the measured R-spectra. For example, KRS-5 window is used as a typical optical window in visible to infrared region but is not available in far-infrared/THz wave region because the onset of the reststrahlen band of the phonon absorption is 25 μm and then not transparent in far-infrared/THz wave region [4]. LiF or quartz is available in visible to ultraviolet, and polyethylene film in far-infrared/THz wave region. We have to replace optical windows of microscope and cryostat according to each spectral region in the R-measurement. Therefore, the best way to exclude such frequent and troublesome replacement of optical windows is to develop a new type of a spectromicroscopic system which has no optical window. (2) In order to study a change in electronic state of solids under high pressure at low temperature, we needs a spectromicroscopic system combined with a DAC. The DAC is required to be equipped with a pressure generation mechanics which is controlled from the outside of the cryostat. However, the installation of such low temperature cryostat-DAC system in conventional microscope is quite difficult because the working distance, the distance between of a pair of Schwarzschild mirrors of the microscope, is usually at longest about 30 mm and then it is very difficult to hold enough space for the installation of the low temperature cryostat and DAC. Although the working distance longer than 80 mm at shortest is required for such an installation, such a microscope with high flexibility in operation has not been commercially available. In order to remove such instrumental difficulties involved in conventional microscopes, it is necessary to develop a microscope which possesses a longer working distance than 80 mm and is equipped with all optical components including of Schwarzchild mirrors in a UHV chamber. These are the reasons why we developed the in-UHV microscope together with low temperature cryostat and DAC.

3.2.2 Specifications of In-UHV Microscope

Schematic drawings of the in-UHV microscope are shown in Fig. 3.2a and the photograph in Fig. 3.2b, respectively. In order to cover a wide wavelength region, two spectrometers were adopted in the microscope. One is a Fourier transformed interferometer, Japan spectroscopy company 660PLUS, equipped with a globar lamp as a light source and two kinds of detectors (mercury-cadmium-telluride (MCT) in near infrared and infrared, and liq. He cooled bolometer in far-infrared/THz wave). The other is a grating monochromator, Jarrel-Ash Com. Monospectro27, equipped with halogen and D-lamps as light sources and photomultiplier as a detector in a visible-ultraviolet region. Visible-ultraviolet signals from samples loaded in the DAC were introduced to a grating monochromator by using of an optical fiberscope. Both light beams, from the near-infrared to far-infrared/THz wave and from the visible to ultraviolet light beam from halogen lamp (or D-lamp) were selected by the switching mirror to the UHV chamber. Thin CVD diamond windows (De Beers Com., IIa type) were used as the inlet and the outlet window of the microscope in order to sep-

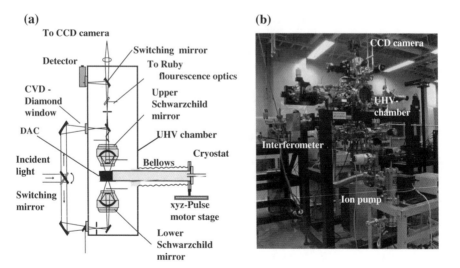

Fig. 3.2 **a** Schematic view of the in-UHV microscope of which all optical components are installed in an ultrahigh vacuum (*UHV*) chamber. Either of a visible to an ultraviolet light or a near-infrared to a far-infrared/THz wave is introduced to the microscope by switching mirror. Optical system to detect visible-ultraviolet signals from sample is not shown. **b** Photograph of the developed in-UHV microscope. The figure is taken from [1] under permission of VSJ

arate the UHV ($\sim 10^{-6}$ Pa) microscope from a low vacuum(a few Pa) interferometer. CVD diamond window of which thickness is 0.25 mm in a central position has a wedge-shape with the taper angle of 1.5° in order to avoid interference fringes due to multireflections within thin film. The multiplication factor of the Schwarzchild mirror of the microscope is 8 and the value of the numerical aperture (NA) is 0.5.

3.2.3 Diamond Anvil Cell

In order to settle a DAC in the narrow space of the microscope, we developed a very compact DAC with the diameter of 12 mm and also a liquid. He flow type of cryostat. The material of a DAC and its lever-arm type of DAC mount is made of Be-Cu with a good thermal conductivity. The DAC is settled at the bottom of the low temperature cryostat which is based on the $x - y - z$ pulse motor stage, and cooled down to 10 K. The generation of high pressures is done by rotating of the screw driver of the cryostat which leads directly to the movement of the lever-arm of the DAC mount from the outside of the cryostat [5, 6]. The cryostat is connected with the UHV chamber of the microscope through a flexible bellows part as shown in the figure. The positioning of the optimal configuration of the cryostat with a DAC is done by the movement of the stage derived by a stepping motor and its accuracy is 1 μm/pulse of the stepping interval of the motor.

3.2.4 Spatial Resolution

In order to know the actual **spatial resolution** of the microscope, we measured the 2D $(X - Y)$ distribution of the reflected light intensities by the sharp edge of gold mirror which was settled on the $X - Y$ scanning stage at the focal position of the microscope. The microscope was operated with the upper aperture with the diameter of 1 mm and the integrated light intensity in the energy range of 700–9000 cm^{-1} was detected by a MCT detector. Measured 2D distribution of the reflected light intensities is shown in Fig. 3.3a, and the differential curve of the intensity distribution curve at the fixed position $Y = 130$ μm in Fig. 3.3b. At the focal position of the microscope the FWHM of the focused image is about 50 μm and the spread of intensity distribution is almost 125 μm which is comparable with the designed spatial resolution determined by the value of NA $= 8$. However, the lights which have longer wavelength components among the integrated light intensity might suffer from the lowering of the resolution due to the diffraction effect.

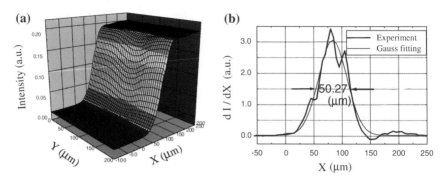

Fig. 3.3 a Two dimensional $X - Y$ distribution of the reflected light intensities by the sharp edge of gold mirror which was settled at the focal position of the microscope. **b** Differential curve of the intensity distribution curve at a fixed position $Y = 130\,\mu m$ in (**a**). *Thick curve* is experimental and thin one is Gaussian fitting curve with the full width at half maximum (*FWHM*) of about $50\,\mu m$, The spread of intensity distribution is almost $125\,\mu m$. The figure is taken from [1] under permission of VSJ

3.2.5 Available Spectral Regions

Diamond window is optically transparent with transmittance of about 70 % from 300 nm to far-infrared/THz wave except the region around $2000\,cm^{-1}$ where the strong absorption occurs due to two phonon absorption in diamond. The adoption of the CVD diamond windows made possible the development of the in-UHV microscope without any other optical windows. The microscope covers the available spectral regions from 300 nm to $100\,\mu m$. Typical example of the measured row spectra are shown in Fig. 3.4. The whole region is composed of five different spectra according

Fig. 3.4 Available spectral regions covered by the in-UHV microscope. Whole region is composed of five different wavelength regions from an ultraviolet to a far-infrared/THz wave which are measurable without breaking of the vacuum of the UHV microscope. Each spectrum is **a** far-infrared/THz wave, **b** infrared, **c** near-infrared, **d** visible, and **e** ultraviolet, respectively. Note the ordinate scale in each curve are not equivalent

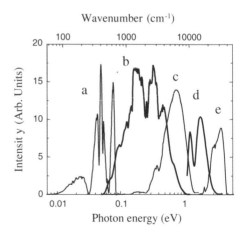

to different light sources, spectrometers and detectors. They are (a) far-infrared/THz wave, (b) infrared, (c) near-infrared, (d) visible, and (e) ultraviolet, respectively.

3.3 Optical Properties of Solids Under Pressure

In this session, we describe the experimental results of spectromicroscopy on ice, $Nd_2Ir_2O_7$ and $Pb_2Ir_2O_7$ as the examples of the successful applications of the in-UHV microscope to the study of optical properties of solids under high pressure at low temperature.

3.3.1 Hydrogen Order-Disorder Transition of Ice

Ice is one of the most familiar materials in our usual life but there exist more than ten kinds of stable phases dependent on temperatures and pressures [7] as shown in Fig. 3.5. We focuss our interest mainly on the optical properties of the two phases, VII and VIII, because they are concerned with hydrogen order-disorder phase transition. The crystal structure of ice VII which belongs to a cubic $Pn3m$ space group [8] is shown in Fig. 3.6a in which open and closed circles indicate the two independent networks structure of the oxygen atoms (denoted as O1 and O2), respectively. Straight bars connecting neighboring oxygen atoms in the same network indicate

Fig. 3.5 Phase diagram of ice versus pressures and temperatures. L corresponds to a liquid phase. The figure is taken from [1] under permission of VSJ

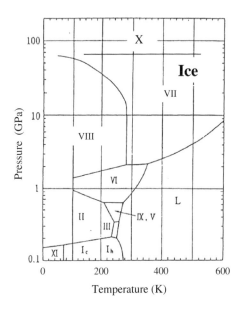

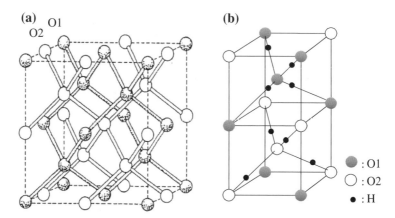

Fig. 3.6 Crystal structure of ice. **a** Two kinds of networks of oxygen atom O1 (*closed circles*) and O2 (*open circles*) in ice VII [7]. **b** The ordered hydrogen atoms (*small closed circles*) at the stable position in the hydrogen-bonding in ice VIII [8]. As the result of the order, the orientation of the dipole moment of the upper cell is antiparallel to that of the lower cell. The figure is taken from [1] under permission of VSJ

the hydrogen bond. We can see there is no hydrogen bond between the different two networks. The hydrogen order-disorder transition originates from the stable two positions for the hydrogen atom in the hydrogen bond. The hydrogen atoms in ice VII possess at random either of the two positions and consequently the orientation of the dipole moment of the water molecule is in at random state (disorder state). However, in ice VIII [9], the hydrogen atoms in the both networks occupy each other the different site so that the orientation of the dipole moments array in the antiferroelectric configuration shown in Fig. 3.6b such as the antiferromagnetic configuration of the spin moment in magnetic materials. The infrared active rotational and translational vibrations of ice VIII have been reported [10, 11]. We follow the same assignments of the observed peaks in the R-spectra.

The transmission measurements on ice VII and VIII under pressures have been performed by Kobayashi et al. [12] on thin films of ice which is loaded in a DAC. They found the sharp absorption around $200\,\mathrm{cm}^{-1}$ at 260 K under 3 GPa assigned to the translational phonon mode $\nu_{T_x,T_y} E_u$ [11] in IceVIII. However, the absorption coefficient was too large to determine the exact phonon peak energy and the FWHM because of the saturation of the top part of the absorption peak. Then, we carried out the R-measurements on polycrystalline ice under high pressures by the spectromicroscopic experiment combined with the DAC in order to determine the exact phonon peak position and its precise pressure-dependence. The R-spectra of ice VII and VIII in the energy regions of 150–$4000\,\mathrm{cm}^{-1}$ are shown in Fig. 3.7. In the spectra of ice VII (300 K, 7 GPa), four peaks were resolved as denoted by ν_R, ν'_R, ν_2 and $\nu_{1,3}$, respectively in Fig. 3.7a. The peak $\nu_{1,3}$ around $3420\,\mathrm{cm}^{-1}$ is the combination of the intramolecular vibration modes ν_1 and ν_3 which are symmetric stretching vibration and asymmetric stretching one. The mode of the intramolecular vibrations ν_1, ν_2, and

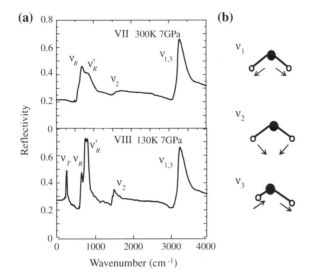

Fig. 3.7 a The R-spectra of ice VII and VIII in the energy regions of 150–4000 cm^{-1}. In the spectra of ice VII (300 K, 7 GPa), four peaks, ν_R, ν'_R, ν_2 and $\nu_{1,3}$, were resolved and in the spectra of ice VIII (130 K, 7 GPa) a new peak was resolved around 265 cm^{-1} which is a translational phonon mode [11]. **b** Schematical drawing of the modes of three intramolecular vibrations

ν_3 are schematically drown in Fig. 3.7b. The doublet peak ν_R around 700 cm^{-1} and around 850 cm^{-1} is corresponding to the $\nu_{R_x,R_y} E_u$ and $\nu'_{R_x,R_y} E_u$ modes [11] which are assigned to the rotation-vibrational modes. The peak ν_2 around 1700 cm^{-1} is the bending vibration mode of the water molecule. In the spectra of ice VIII (130 K, 7 GPa), a new peak ν_T was resolved around 265 cm^{-1} which is the translational phonon mode $\nu_{T_x,T_y} E_u$ and corresponds to the strong absorption band in the transmission measurements.

The evolution of the peak ν_T of ice VIII at 130 K with pressures of (a) 4.5, (b) 8.8, and (c) 11.0 GPa is shown in Fig. 3.8a together with the spectrum of ice VII at 300 K under 5.5 GPa which is shown by a broken line in the figure. In ice VII, the spectrum has no distinct peak structure but is almost flat which is determined by the refractive index. In ice VIII, on the other hand, the sharp peak with the FWHM of 30 cm^{-1} was resolved around 235 cm^{-1} at 130 K under 4 GPa due to the ordering of the hydrogen bond. Such small FWHM value indicates the small energy difference between the transverse and longitudinal phonon in ice VIII. The observed peak energies $\nu(P)$ in the R-spectra with pressures increased up to 300 cm^{-1} under 11.0 GPa and they were plotted in Fig. 3.8b. The value of the **mode Gruneisen parameter** γ_i which is given by $\gamma_i = -(\frac{d \ln \nu_i}{d \ln V})$ was determined to be 1.74 for the mode $\nu_{T_x,T_y} E_u$ by using of the observed $\nu(P)$ values and the lattice compressibility data of ice VIII.

The evolution of the spectra of ice VIII at 130 K with pressures is shown in Fig. 3.9. The change in the peak energies of the ν_R mode with pressure is 19.0 cm^{-1}/GPa in the pressure range 6–9.8 GPa which is almost same with ν'_R mode. The values of the

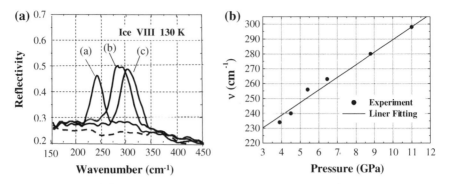

Fig. 3.8 a Evolution of the peak of polycrystalline ice VIII at 130 K with pressures of (*a*) 4.5, (*b*) 8.8, and (*c*) 11.0 GPa, respectively. **b** The peak shifts of the mode with pressures. From the peak energy shifts and the volume compressibility, the mode Gruneisen parameter was estimated to be 1.54

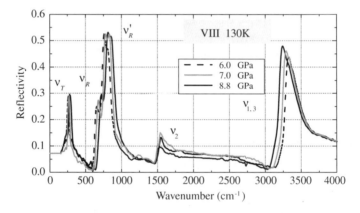

Fig. 3.9 Evolution of the spectra of ice VIII with pressures at 130K. From the peak energy shifts with pressures, the mode Gruneisen parameter was estimated to be 1.23 for the bands ν_R and ν'_R

mode Gruneisen parameter were determined to be 1.23 for the ν_R and ν'_R modes by using of the observed $\nu(P)$ values and the same lattice compressibility data of ice VIII.

On the order-disorder transition of ice under pressures, ice is known to take place a further transition from ice VII to ice X around 60 GPa, which is called as the hydrogen bond symmetrization [13]. Hydrogen bond symmetrization means that the two stable positions for hydrogen atoms of ice VII approach each other by high compression and finally hydrogen atoms possess the middle position in the hydrogen bond between the neighboring oxygen atoms. Then, the oxygen atoms form a body-centered cubic structure. Hitherto, considerable interest has been focused on the presence of symmetric ice X and its physical properties because ice has been considered as the prototype of the hydrogen-bonded molecular materials. Infrared

Fig. 3.10 Evolution of the peak in the R-spectra round at $3420\,\mathrm{cm}^{-1}$ with pressures at room temperature which is assigned to the intramolecular vibrational mode. We can see the red shifts of the peak with pressures which means the instability of the hydrogen bond toward the symmetric ice X

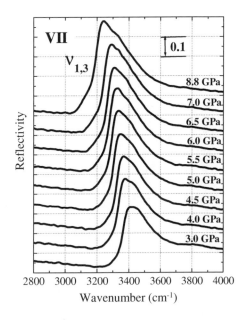

measurements on ice X have been performed by Aoki et al. [14] and Goncharov et al. [15]. They reported the evolution of the infrared spectra with pressures indicating the softening of the energy of the intramolecular vibrational $\nu_{1,3}$ mode and the hydrogen bond symmetrization above 60 GPa. We carried out the R-measurements on ice VII under pressures up to 8.8 GPa at room temperature. The evolution of the $\nu_{1,3}$ peak of ice VII is shown in Fig. 3.10. The $\nu_{1,3}$ mode showed the gradual decrease in the peak energies (softening) with pressures. The amount of the change in the energies of the mode with pressures is $-30\,\mathrm{cm}^{-1}$/GPa in the pressure range of 3–8.8 GPa which is in good agreement with the value reported by Aoki et al. The present data confirms that from the beginning of the ice VII even in low pressure region the intramolecular vibrational mode undergoes the softening towards the symmetric phase ice X.

3.3.2 $Nd_2Ir_2O_7$

$Nd_2Ir_2O_7$ is one of $Ln_2Ir_2O_7$ family of **Ir-based pyrochlore oxides**. The fundamental constitution is represented as $A_2B_2O_7$ where A is rare earth element (Ln^{3+}) and B is transition metal (Ir^{4+}). A(B) forms each other independent sublattice network of corner-sharing tetrahedra O-A$_4$ (O-B$_4$) in pyrochlore structure with a cubic space group $Fd\bar{3}m$ [16] in which oxygen atom locates at the center of the tetrahedra. Ir atom locates also at the central site of octahedra Ir-O$_6$ and Ir-5d electrons undergo a crystal field splitting into e_g and t_{2g} states. Ir-5$d_{t_{2g}}$ electrons mix with conduction electrons and contribute to electronic transport properties. On the other hand, Ln-$4f$ electrons are well localized and do not contribute to electronic transport properties, because they locate by several eV in energy below the Fermi level. Both Ln-$4f$ and Ir-$5d$ electrons contribute to the magnetism of the compound. Recently, $Ln_2Ir_2O_7$ has attracted considerable interest in solids state physics because they show very attractive phenomena such as a geometrical 3D magnetic spin fluctuation essentially due to its pyrochlore structure, temperature-induced metal-insulator (M-I) phase transition, and so on. $Ln_2Ir_2O_7$ compounds exhibit a second order of electronic phase transition from a metallic state with a paramagnetism (PM) at high temperatures to an insulating phase with an antiferromagnetism (AF) at low temperature as shown in Fig. 3.11 [2]. Vertical upward arrows indicate the ionic radii of Ln ions. The critical temperature of the phase change, denoted as T_{MI}, of $Nd_2Ir_2O_7$, and $Sm_2Ir_2O_7$ are 37 and 117 K, respectively. We can see that the values of the T_{MI} increase with increasing of atomic numbers of Ln ions. In rare earth compounds, a lattice contraction with an increase in atomic number of rare earth elements is known as a so-called **lanthanide contraction** effect[17] that means the decrease in the ionic radii of Ln ions with the atomic number which result in the decrease in the lattice constant of the compounds. Among $Ln_2Ir_2O_7$, $Pr_2Ir_2O_7$ which possesses the largest ionic radius does not undergo a phase transition and stay always metallic in a whole temperature region.

Spectroscopic studies to investigate the change in the electronic states have not yet been done on $Ln_2Ir_2O_7$ compounds except experiments on macroscopic physical quantities such as electric resistivity, heat capacity, and magnetic susceptibility. Then, we performed the R-measurements on $Nd_2Ir_2O_7$ at 80 K under high pressure up to 10 GPa in order to study the change in the electronic states due to the phase transition. The evolution of the R-spectra of $Nd_2Ir_2O_7$ at ambient pressure is shown in Fig. 3.12. The reflectivity of the low energy part below $1000\,cm^{-1}$ in the R-spectra at temperatures above 40 K is very high and $R \leq 1$ due to the existence of the conduction electrons (**Drude reflection**). However, many sharp lines

Fig. 3.11 Phase diagram of $Ln_2Ir_2O_7$ (Ln = rare earth element) plotting of the T_{MI} with ionic radii of Ln^{3+} ions. Upward arrows are the ionic radius of Pr, Nd, Sm, Eu, and Gd, respectively. *Vertical downward arrow* $\overrightarrow{AB}$ indicates the cooling process for $Nd_2Ir_2O_7$ at ambient pressure, and *horizontal broken arrow line* $\overrightarrow{AC}$ indicates the lattice compression process at 80 K

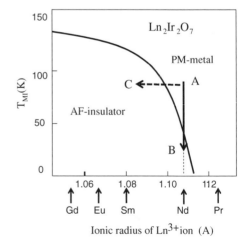

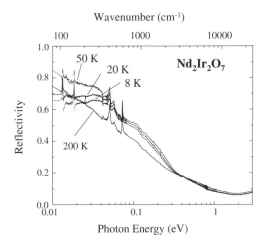

Fig. 3.12 R-spectra of $Nd_2Ir_2O_7$ at different temperatures at ambient pressure

appeared in the spectra below $1000\,cm^{-1}$ superimposed on the metallic reflectivity. They come from the phonon peaks caused by the weak screening effect of the phonon polarization by the low density of the conduction electrons. We can see that the R-spectra below $500\,cm^{-1}$ gradually lose its intensities with cooling below $40\,K$. This phenomena correspond to the change of the metallic electronic state to the insulating state. In general, in a change in optical spectra from a metallic state to an insulating state, a new peak due to an interband transition appears instead of a suppression of a Drude reflection due to a drastic decrease in a density of conduction electrons. In the case of $Nd_2Ir_2O_7$, a new peak was found to grow around $1500–2000\,cm^{-1}$ indicating of the formation of the energy band gap in the insulating states. Among many phonon peaks found in $300–600\,cm^{-1}$ regions of the R-spectra, the peak energy of the phonon at $585.6\,cm^{-1}$ in Fig. 3.12 was plotted as the representative with temperatures in Fig. 3.13a. The phonon at $585.6\,cm^{-1}$ is the phonon mode concerned with the vibrational mode to change the Ir-O-Ir bond angle and the bond length in the pyrochlore structure. In general, a usual phonon mode shows a blue shift on peak position with a cooling due to a suppression of an anharmonicity in phonon-phonon interaction at low temperature. In a cooling process, the phonon at $585.6\,cm^{-1}$ showed a blue shift in the metallic temperature regions as usual, but in the insulating phase (below the T_{MI}), showed gradually a red shift (**softening of phonon**) against the expected curve for the usual phonon mode as shown by a broken line in the figure. Observed phonon softening suggests the occurrence of the instability of the crystal structure in connection with the phase change. Now, from the observed red shift we may consider that the modification of the lattice structure (the lowering of the symmetry) occurs in the phase transition to the insulator.

The phase diagram indicates that the T_{MI} increases with the decrease in the ionic radii of Ln ions. However, on the other hand, the temperature-dependent R-spectra of $Nd_2Ir_2O_7$ (see Fig. 3.13a) suggests the occurrence of the lattice modification on the phase change. In order to examine if the lattice contraction may play an important role

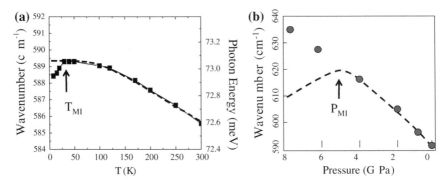

Fig. 3.13 **a** Temperature dependence of the phonon peak energy of $Nd_2Ir_2O_7$ at 585.6 cm^{-1} at room temperature and at ambient pressure in Fig. 3.12. *Broken curve* is the expected one to the conventional phonon mode. We can see the phonon softening below the T_{MI}. **b** Evolution of the phonon peak energies with different pressures at 80 K. The *broken curve* indicates a schematic drawing for the possible phonon softening according to the expected pressure-induced phase transition in the lattice compression process shown by a *horizontal broken line* $\overrightarrow{AC}$ in Fig. 3.11. P_{MI} means the critical pressure which may induce the phase transition from the PM-metal to the AF-insulating phase by the lattice compression. See the text for the details

in the M-I transition or not, we carried out R-measurements by the spectromicroscopy on $Nd_2Ir_2O_7$ under high pressure up to 10 GPa. If the lanthanide contraction works as the cause of the M-I transition, uniform lattice compression by application of high pressure to samples will induce the similar phonon softening with the T-dependent R-spectra at ambient pressure. The lattice compressibility data of $Nd_2Ir_2O_7$ has not yet been reported, but the value of -1.63 Å/GPa is reported for $Nd_2Mo_2O_7$ [18] which belongs to the similar pyrochlore oxides with $Nd_2Ir_2O_7$. Assuming the same value for $Nd_2Ir_2O_7$, the pressure of 5 GPa can compress the lattice constant of $Nd_2Ir_2O_7$ to 10.302 Å which is smaller than 10.318 Å of $Sm_2Ir_2O_7$. Considering that the T_{MI} of $Sm_2Ir_2O_7$ is 117 K, the pressure of 5 GPa at 80 K is enough to lead $Nd_2Mo_2O_7$ into the insulating phase along the horizontal broken line AC as shown in Fig. 3.11. Then, we assume tentatively 5 GPa as the critical pressure P_{MI} to realize the pressure-induced M-I transition for $Nd_2Ir_2O_7$ if the lattice contraction works as the driving force of the M-I transition. The peak shifts of the 585.6 cm^{-1} band at ambient pressure were plotted with pressures in Fig. 3.13b. The broken curve in the figure indicates the expected one for the phonon softening with pressures above the critical pressure, P_{MI} of 5 GPa. Experimental results indicated the increase in the phonon peak almost linearly with pressures over all pressure regions up to 10 GPa and the M–I transition does not occur. This result seems to suggest that the main cause of the phase transition is not the lattice contraction but the lattice modification due to the change in the Ir-O-Ir bond angle and the bond length. This suggestion seems to be reasonable if we consider that the increase in the mixing of Ir-$5d_{t_{2g}}$ state with conduction electrons by the change in lattice modification affects the transport properties and uniform lattice compression does not affect the M–I transition. This prediction on the origin of the

M–I transition will be verified by precise X-ray diffraction experiments to detect such a slight lattice modification by using of a high brilliant X-ray source such as SPring-8 which is the electron storage ring facilities at Hyogo Prefecture (Japan) and the most powerful synchrotron light source to supply with a sharp X-ray beam [19].

3.3.3 $Pb_2Ir_2O_7$

Just like $Nd_2Ir_2O_7$, $Pb_2Ir_2O_7$ is one of the Ir-based pyrochlore oxides of which magnetic Ln ions are substituted by nonmagnetic Pb element. As mentioned at the previous session, $Ln_2Ir_2O_7$ compounds shows very attractive phenomena such as a geometrical 3D magnetic spin fluctuation and temperature-induced M-I transition. Such properties depend strongly on the kind of Ln ions because the interaction between Ln^{3+}-$4f$ and Ir-$5d$ electrons which locate below the conduction bands changes with the kind of Ln ions. Then, $Pb_2Ir_2O_7$ was considered to be the most important compound as the reference materials because Pb ion is nonmagnetic and then $Pb_2Ir_2O_7$ works as the litmus paper to examine how Ln^{3+}-$4f$ electrons contribute to the M-I transition and the magnetism, for example. Recently, a very minute single crystal of $Pb_2Ir_2O_7$ has been successfully synthesized. However, the physical and chemical properties have not yet been investigated except the transport data [20]. According to the electrical resistivity data, $Pb_2Ir_2O_7$ exhibits metallic properties at all temperature regions at ambient pressure. First, we measured the temperature dependence of the R-spectra at ambient pressure. Photograph of synthesized single crystal of $Pb_2Ir_2O_7$. is shown in Fig. 3.14a and the evolution of the R-spectra with temperature at ambient pressure in Fig. 3.14b. Even at low temperature, the Drude type of reflectivity was observed and the overall spectral profile does not change from the room temperature. This means that the density of the conduction electrons and the effective electron mass almost do not change and $Pb_2Ir_2O_7$ stays metallic at all temperature region as transport data suggested.

Next, we carried out the spectromicroscopic measurements in order to investigate the change in the electronic states with pressure. The evolution of the R-spectra with pressures at 40 K was shown in Fig. 3.15. The metallic reflectivity at ambient pressure changes gradually with pressure and at 10 GPa the clear dip structure was found to grow in the spectral region below 1 eV which corresponds to the formation of the energy gap across the Fermi energy level. That is, the present spectromicroscopic measurements under pressure revealed (1) the formation of the **Pseudo energy gap** due to the pressure-induced phase transition from metallic states to insulating states, (2) the magnetic moments of the $4f$-electrons do not affect at least the occurrence of the phase transition in the Ir-based oxides, and (3) $Pb_2Ir_2O_7$ exhibits the change from metallic states to insulating states under pressure. The observed M-I process in $Pb_2Ir_2O_7$ is the inverse to the normal process because conventional materials change from insulating states to metallic ones under pressure. Similar such inverse process under pressure has been reported in Cu spinel compound [21]. We expect that the

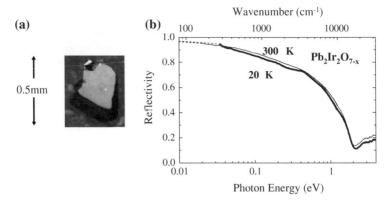

Fig. 3.14 **a** Photograph of synthesized single crystal of $Pb_2Ir_2O_7$. **b** Temperature dependence of the R-spectra of $Pb_2Ir_2O_7$ at ambient pressure. The figure is taken from [1] under permission of VSJ

Fig. 3.15 Evolution of the R-spectra of $Pb_2Ir_2O_7$ at ambient pressure, 5 and 10 GPa at 40 K. The figure is taken from [1] under permission of VSJ

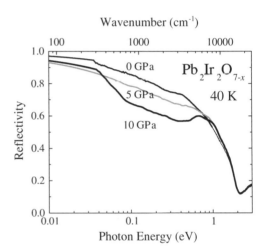

clear occurrence of the **pressure-induced M-I transition** in $Pb_2Ir_2O_7$ might be proved by the other various kinds of high pressure experiments in the close future.

3.4 Conclusions

In this article, we described the details of the development of the in-UHV microscope which is available in the wide wavelength regions from ultraviolet to far-infrared/THz wave as the tool of the spectromicroscopy to investigate the change in the electronic state of solids under high pressure. The results on ice, $Nd_2Ir_2O_7$, and $Pb_2Ir_2O_7$ were reported as the successful examples of the applications of the in-UHV microscope.

Obtained spatial resolution of the microscope was about $50\,\mu$m at the FWHM at the focal position. Nowadays, the application of the spectromicroscopic measurement is required particularly in the far-infrared/THz wave region in various fields of science, for example, studies of a collective motion in soft matters, physical properties of solids under experimental multiextreme conditions of pressure, low temperature and high magnetic field in order to obtain informations on electronic states of solids under high pressure. However, the spectroscopic experiments under high pressure in the far-infrared/THz wave region is quite rare because of the too weak intensity of conventional light sources to obtain enough signals from very small-area samples loaded in a DAC. In order to obtain the better spatial resolution and to overcome the experimental difficulties in the performance of the spectromicroscopic measurements in the far-infrared/THz wave region, infrared synchrotron radiation light should be adopted as the most brilliant light source instead of conventional black body sources. For example, the infrared beamline which is mainly dedicated to the spectromicroscopic experiments on 2D imaging spectroscopy and spectroscopy under multiextreme experimental conditions has been developed at SPring-8 [22], and at present works in visible to far-infrared/THz wave region.

Acknowledgments We are grateful to Prof. K. Matsuhira (Kyushu Institute of Technology) for supplementing with $Nd_2Ir_2O_7$ and $Pb_2Ir_2O_7$ samples and helpful discussions, and also to Mr. H. Yamano and Mr. M. Nishiyama (Kobe University) for their helps in the performance of high pressure experiments.

References

1. T. Nanba, J. Vac. Soc. Jpn. **53**, 406 (2010). (in Japanese)
2. K. Matsuhira, M. Wakeshima, R. Nakanishi, T. Yamada, A. Nakamura, W. Kawano, S. Takagi, Y. Hinatsu, J. Phys. Soc. Jpn. **76**, 043706 (2007)
3. F. Wooten, *in Optical Properties of Solids* (Academic Press, New York, 1972)
4. D.E. McCarthy, Appl. Optics **6**, 1896 (1967)
5. I.F. Silvera, R.J. Wijngaarden, Rev. Sci. Instrum. **56**, 121 (1985)
6. T. Nanba, Rev. Sci. Instrum. **60**, 1680 (1989)
7. Ph Pruzan, J. Mol. Struct. **322**, 279 (1994)
8. J.D. Jorgensen, T.G. Worlton, J. Chem. Phys. **83**, 329 (1985)
9. A.J. Brown, E. Whalley, J. Chem. Phys. **45**, 4360 (1966)
10. E. Walley, Can. J. Chem. **55**, 3429 (1977)
11. S.P. Tay, D.D. Klug, E. Whalley, J. Chem. Phys. **83**, 2708 (1985)
12. M. Kobayashi, T. Nanba, M. Kamada, S. Endoh, J. Phys. Condens. Matter **10**, 11555 (1998)
13. K.S. Schweizer, F.H. Stillimger, J. Chem. Phys. **80**, 1230 (1984)
14. K. Aoki, H. Yamawaki, M. Sakashita, H. Fujihisa, Phys. Rev. B **54**, 15673 (1996)
15. A.F. Goncharov, V.V. Struskin, M.S. Somayazulu, R.J. Hemley, H.K. Mao, Science **273**, 218 (1996)
16. J.N. Millican, R.T. Macaluso, S. Nakatsuji, Y. Machida, Y. Maeno, J.Y. Chan, Mater. Res. Bull. **42**, 928 (2007)
17. R.D. Shannon, Acta Crystallogr. Sect. A **32**, 751 (1976)
18. H. Ishikawa, S. Xu, Y. Morimoto, A. Nakamura, Y. Ohishi, Y. Mizutani, Phys. Rev. B **70**, 104103 (2004)

19. http://www.spring8.or.jp/
20. K. Matsuhira, (Kyushu Institute of Technology), private communication.
21. L. Chen, M. Matsunami, T. Nanba, T. Matsumoto, S. Nagata, Y. Ikemoto, T. Moriwaki, T. Hirono, H. Kimura, J. Phys. Soc. Jpn. **74**, 1099 (2005)
22. S.Kimura et al., Nucl. Instr. and Meth. A 467–468, 893(2001). See also the page of BL43IR in the Beamline list in [19].

Chapter 4
Infrared and Terahertz Synchrotron Radiation: Optics and Applications

Shin-ichi Kimura

4.1 Introduction

One method for creating intense as well as bright terahertz (THz) and infrared (IR) light is by using radiation from relativistically accelerated electrons, such as synchrotron radiationSR and free-electron laserFEL. SR is the radiation emitted from bending magnets and insertion devices (undulators/wigglers). Because SR consists of broadband electromagnetic waves from the X-ray to the THz range, light with the desired wavelength and intensity (photon flux) can be obtained by using suitable optics. SR is a good source for spectroscopy, and over 20 SR facilities worldwide are equipped with IR/THz beamlines. In contrast, FEL is monochromatic light with high coherence and relatively high intensity. These properties of FEL make it worthwhile as an excitation source and as monochromatic probing light.

The size of the SR light source is determined by the size of the electron beam in the accelerator. The electron beam size is similar to the wavelength of THz light (i.e., sub millimeter size). Because the emission angle of SR is small, the electron beam emittance (i.e., the area occupied by the particles of the beam in space and momentum phase space) is smaller than that of the THz light. The SR light can, therefore, be focused down to spot size, with an ideal diffraction limit. This property makes the light suitable for microscopic spectroscopy (micro-spectroscopy). In fact, the purpose of almost all of the world's IR/THz beamlines is microspectroscopy and microimaging with higher spatial resolution than with conventional IR microscope.

In the case of SR light sources, the emission point is located in a vacuum chamber with a bending magnet. Since the vertical space available for collecting the SR is narrow, all of the emitted light cannot be collected. If we have a very wide horizontal acceptance angle, the emission point becomes an emission arc. Therefore, laboratory optics cannot be utilized with IR/THz SR. In addition, the emission point in an

S. Kimura (✉)
Graduate School of Frontier Biosciences, Osaka University, 1-3 Yamadaoka, Suita,
Osaka 565-0871, Japan
e-mail: kimura@fbs.osaka-u.ac.jp

K. Shudo et al. (eds.), *Frontiers in Optical Methods*,
Springer Series in Optical Sciences 180, DOI: 10.1007/978-3-642-40594-5_4,
© Springer-Verlag Berlin Heidelberg 2014

electron accelerator is in an ultra-high vacuum ($\sim 10^{-8}$ Pa), whereas THz light needs a pressure of less than 1 Pa for excluding the absorption of water vapor. To separate the ultra-high vacuum of the accelerator and the low vacuum of the downstream optics, optical windows are used. Because SR light is more broadband than that from conventional THz/IR light sources, the use of optical windows restricts the wavelength regions available. To optimize the use of the wide wavelength region of SR, exchange of optical windows without breaking the vacuum is effective.

In Japan, there are currently three IR/THz beamlines. The first IR/THz beamline, BL6A1, has been constructed at UVSOR in 1986 [1]. In 2003, UVSOR was upgraded to the lowest emittance ring (27 nm rad) among small SRs, with an acceleration energy below 1 GeV and its name was changed to UVSOR-II [2]. Simultaneously, BL6A1 was upgraded to the new IR/THz beamline, BL6B, with high brilliance and high flux [3]. To obtain higher performance, we employed a so-called "magic mirror" with a large acceptance angle (215(H) × 80(V) mrad^2) [4]. Upgrading the beamline made the photon flux of BL6B four times larger, and the brilliance 10^2 times larger, than those of BL6A1. In 1999, another IR beamline was constructed at SPring-8 (in Hyogo Prefecture, Japan) [5], which is the largest SR ring in the world. The beamline was equipped with a three-dimensional magic mirror for the first time. At SPring-8, spectroscopy at high temperature and high pressure are used to investigate conditions inside the Earth. Spectroscopy of correlated materials at low temperature and high pressure is also performed [6]. In 2009, a new IR microspectroscopy beamline was constructed at the small SR ring, AURORA, at Ritsumeikan University [7]. The beamline will be dedicated to industrial use.

To perform IR/THz spectroscopy using SR, special optics that differ from those used in laboratory systems are needed. In this chapter, the IR/THz optics developed for use with SR are introduced, as in an example of their application using IR/THz SR.

4.2 Principle of Three-Dimensional Magic Mirror Optics for THz/IR SR

SR is characterized by the emission angle of light. When a relativistic electron beam moves in an external field, the radiation spectrum ($\partial^2 P(\omega)/\partial\omega\partial\Omega$) from a minute orbital is [8]

$$\frac{\partial^2 P(\omega)}{\partial\omega\partial\Omega} = \frac{e^2}{12\pi^3\varepsilon_0 c}\left(\frac{\omega\rho}{c}\right)^2\left(\frac{1}{\gamma}+\theta^2\right)\times\left\{K_{2/3}^2(\xi)+\frac{\gamma^2\theta^2}{1+\gamma^2\theta^2}K_{1/3}^2(\xi)\right\} \quad (4.1)$$

Here, e is the elemental charge, ε_0 the dielectric constant of the vacuum, c the velocity of light, ρ the orbital radius, and γ the ratio of energy to rest mass. $K_{2/3}(\xi)$ and $K_{1/3}(\xi)$ are modified Bessel functions, and $\xi = \omega R/3c(1/\gamma^2 + \theta^2)^{3/2}$, where R is the radius of the electron orbit. The function corresponds to the vertical distribution of SR in the orbital plane. For instance, the vertical emission of the bending-magnet

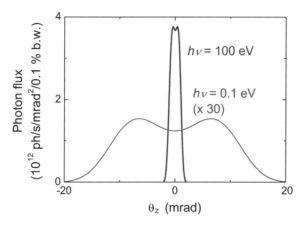

Fig. 4.1 Vertical distribution of bending-magnet radiation of the UVSOR-II storage ring at photon energies of 0.1 and 100 eV

radiation at UVSOR-II is shown in Fig. 4.1. In the figure, the emission at a photon energy of 100 eV is located around the orbital plane ($\theta_z = 0$ mrad), but the emission of the 0.1 eV photon energy expands above 10 mrad from the orbital plane. The intensity at 0.1 eV is much lower than that at 100 eV. These properties imply that the vertically wide acceptance angle at 0.1 eV is needed for high intensity.

On the horizontal axis, on the other hand, the emission from a minute emission angle is the same as in the above function, but the circular orbit of electron beams made by the Lorentz force due to the presence of the dipole magnet is important. This indicates that SR from the dipole magnet is not an ideal point source, but is instead an arc source. The light cannot be focused to a point by using circular/toroidal mirrors. Thus, these mirrors cannot maintain high brilliance.

The "magic mirror" for bending-magnet radiation focused to one point was discovered by Lopez-Delgado and Szwarc in 1976 [9]. In this paper, the shape was defined only in the orbital plane. In reality, however, SR is emitted not only in the orbital plane but also to the horizontal axis. We therefore extended the shape of the magic mirror to a three-dimensional one, creating a "three-dimensional magic mirror (3D-MM)," and we installed this mirror in the IR beamline (BL43IR) at SPring-8 [10]. SPring-8 is the largest SR ring in the world with an orbital radius of about 40 m. The acceptance angle of 36 mrad corresponds to the emission length of 1.44 m (Fig. 4.2a). If a toroidal mirror is added to collect the light, then the focal image, calculated by a **ray tracing method**, is not good (Fig. 4.3a). This result indicates that the high brilliance of SR cannot be preserved by using a toroidal mirror. We therefore employed the 3D-MM. The following calculations are used to form the 3D-MM:

$$\overline{AM}(\theta) = \frac{\frac{1}{2}\{(d_0 - \rho\theta)^2 - \rho^2 a^2\} + a\rho\sin\theta}{d_0 - \rho\theta - a\cos\theta} \tag{4.2}$$

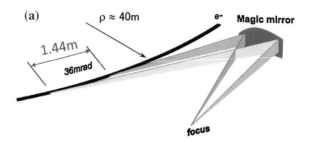

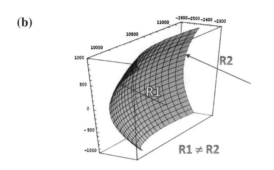

Fig. 4.2 Schematic figure of the optics using a three-dimensional magic mirror (3D-MM) at BL43IR of SPring-8 (**a**) and the curvature of the 3D-MM surface (**b**)

$$\overline{MI}(\theta) = \sqrt{\rho^2 + \overline{AM}(\theta)^2 + a^2 - 2a(\rho\sin\theta + \overline{AM}(\theta)\cos\theta)} \tag{4.3}$$

$$R(\theta) = \frac{2\overline{AM})(\theta)(d_0 - \rho\theta - \overline{AM}(\theta)}{d_0 - \rho\theta} \times \cos\left\{\frac{1}{2}\cos^{-1}\left(\frac{\overline{AM}(\theta) - a\cos\theta}{\overline{MI}(\theta)}\right)\right\} \tag{4.4}$$

$$x(\theta, \nu) = \rho\cos\theta - \overline{AM}(\theta)\sin\theta + \frac{|R(\theta)|\{\sin\theta - \frac{\rho\cos\theta - \overline{AM}(\theta)\sin\theta}{\overline{MI}(\theta)}}{\sqrt{2 + \frac{2(\overline{AM}(\theta) - a\cos\theta)}{\overline{MI}(\theta)}}}\left\{1 - \cos\left(\tan^{-1}\frac{\nu}{R(\theta)}\right)\right\} \tag{4.5}$$

$$y(\theta, \nu) = \rho\sin\theta + \overline{AM}(\theta)\cos\theta + \frac{|R(\theta)|\{-\sin\theta + \frac{a - \rho\sin\theta - \overline{AM}(\theta)\cos\theta}{\overline{MI}(\theta)}}{\sqrt{2 + \frac{2(\overline{AM}(\theta) - a\cos\theta)}{\overline{MI}(\theta)}}}\left\{1 - \cos\left(\tan^{-1}\frac{\nu}{R(\theta)}\right)\right\} \tag{4.6}$$

$$z(\theta, \nu) = R(\theta)\sin\left(\tan^{-1}\frac{\nu}{R(\theta)}\right) \tag{4.7}$$

Here, ρ is the orbital radius of electron beams, a the distance between the center of the electron beam and the focal position, d_0 the optical length from the standard

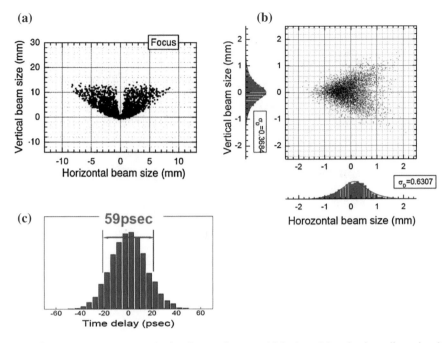

Fig. 4.3 Ray tracing images at the first focus using a toroidal mirror (**a**) and a three-dimensional magic mirror (**b**) at BL43IR of SPring-8. **c** Time structure at the focal point, using the three-dimensional magic mirror

position, θ the acceptance angle of SR, and ν the vertical size of the magic mirror. For instance, the parameters for BL43IR at SPring-8 are as follows: $\rho = 39271.8$ mm, $a = 46025.4$ mm, $d_0 = 69807.4$ mm, $\theta = 1.4804$ to 1.5169 rad, $\nu = -22$ to $+22$ mm. The meanings of $\overline{AM}$, $\overline{MI}$, and R are the same as in [9]. The vertical radius of the magic mirror is obtained for an approximation of a spherical shape by using the distance from the emission point to the focal point.

The shape of a magic mirror surface with the above parameters is shown in Fig. 4.2b. The surface is very complex because of the nonspherical and nonsymmetric shape. To fabricate the surface, a recently developed numerical cutting method is used.

The focal image calculated by using the ray tracing method is shown in Fig. 4.3b. The focal size is much smaller than that with the toroidal mirror (see Fig. 4.3a). The magic mirror cannot only focus SR from the circular electron orbital; it can also reduce the time structure of the emission length to the longitudinal width of the electron bunch (~100 ps), because the magic mirror is characterized by a constant distance from the emission point to the focal point including the trajectory of the electron bunch. The calculated time structure is shown in Fig. 4.3c. In the calculation, the beam length was assumed to be 15 ps. The time structure at the focal point was evaluated to be 59 ps about four times longer than the electron beam length.

Fig. 4.4 Overview of the UVSOR-II storage ring

4.3 IR/THz Beamline Optics at UVSOR-II

The 3D-MM was installed in the THz/IR beamline BL6B at UVSOR-II (Fig. 4.4) in 2004 [3]. UVSOR-II is a relatively small SR ring with a circumference of about 53 m and an electron beam bending radius is 2.2 m. If a 3D-MM is introduced into such a small SR ring, a large acceptance angle can be employed. For instance, at UVSOR-II, the acceptance angle became 215 mrad in the horizontal plane and 80 mrad in the vertical plane. The combination of a large acceptance angle and the shape of the 3D-MM gives high brilliance.

The layout and optics of the beam extraction part (front-end) are shown in Fig. 4.5a and b, respectively. The first mirror in the bending-magnet chamber is a magic mirror with dimensions of 300 mm (horizontal) by 100 mm (vertical). A copper pipe (diameter 5 mm) for water cooling is located in the orbital plane in front of the magic mirror to reduce heat load due to the increased photon energy of the SR.

The THz/IR SR extracted by the magic mirror is derived to the first focusing point (P1) by two plane mirrors (M1 and M2). The position and angle of the light can be controlled by M1 and M2 (Fig. 4.5b). The light emittance of the THz/IR beamline is the product of the light size and the emission angle at P1. The beam size at P1 calculated by the ray-racing method, and the actual beam size obtained, are shown in Fig. 4.6a and b. The beam in the experimental result is spherical and about 1.2 mm in diameter, in contrast with the elliptical shape of 1×2 mm^2 in the calculation. This is because the calculation was done at a photon energy of 0.1 eV, whereas the experimental result was obtained at a photon energy range of 0.05–2 eV by using an HgCdTe (MCT) detector. The result indicates that the SR light can be ideally focused by using the 3D-MM.

The distance from the magic mirror to P1 is 2.5 m and the horizontal solid angle from P1 to the magic mirror is about 86 mrad. By using the experimental beam

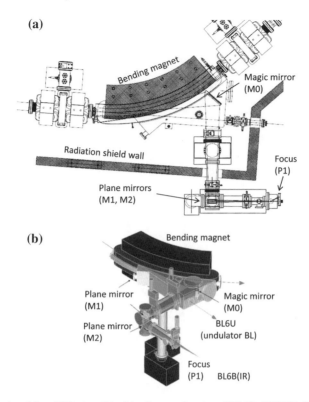

Fig. 4.5 Top view (**a**) and 3D view (**b**) of the front-end optics of BL6B, UVSOR-II

size at P1, the horizontal **emittance** at a photon energy of 0.1 eV ($\sigma_x \cdot \sigma'_x$) becomes 52 μm·rad. In the vertical plane, on the other hand, the emission angle ($\sigma'_y \sim 12$ mrad) shown in Fig. 4.1 at 0.1 eV and the lengths between the center of the emission arc and the magic mirror (about 1 m) and between the magic mirror and P1 (about 2.5 m) make the emittance ($\sigma_y \cdot \sigma'_y$) 5.4 μm·rad. The emittance of a conventional IR spectrometer is $\sigma_{x,y} \cdot \sigma'_{x,y} = 200$–1000 μm·rad. This implies that SR light has a much lower emittance than conventional laboratory sources.

This low emittance gives high brilliance. In fact, in terms of black body emission at 1400 K, the brilliance of BL6B at UVSOR-II is several orders of magnitude greater than that of a conventional source (Fig. 4.7). The difference in the THz photon energy region below 50 meV is much larger than that in the IR region above 50 meV. This indicates that the light is suitable for microspectroscopy in the THz region.

4.4 IR/THz Beamline Optics at UVSOR-II

The BL6B IR/THz of UVSOR-II covers the range of 0.5 meV to 2 eV (4–16,000 cm^{-1}) by using two **Fourier transform IR interferometers (FTIRs)**. One is a **Martin-Puplett-type (MP)** interferometer (FARIS-1, Jasco Inc.) that covers a

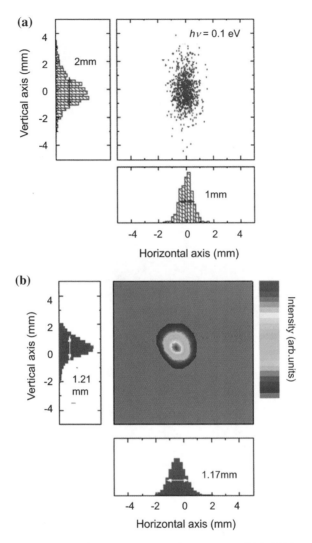

Fig. 4.6 Ray trace results at the first focal point at a photon energy of 0.1 eV (**a**) and beam profile measured by using an MCT detector (**b**) at the first focusing point (P1) in Fig. 4.5

photon energy range of 0.5–10 meV. The other is **Michelson-type interferometer (MI)** (Vertex 70v, Bruker Inc.) covering 5 meV to 2 eV and with four beam splitters (Mylar 23 μm, Mylar 6 μm with multilayer coating, Ge/KBr, and quartz). The beamline layout is shown in Fig. 4.8. The IR/THz SR focused by the 3D-MM is formed into a parallel beam by a collimator (M4) and is guided to the MP. When the parabolic mirror M3 is placed in the optical path, the IR/THz SR is guided to the MI. The IR/THz SR passing through the MP is guided to the reflection/absorption

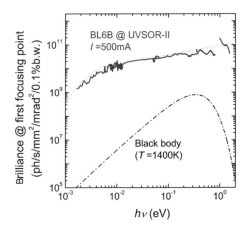

Fig. 4.7 Calculated (*dashed line*) and experimentally obtained (*solid line*) brilliance of BL6B, UVSOR-II compared with those of a thermal (black body) light source at $T = 1400$ K (*dash-dotted line*)

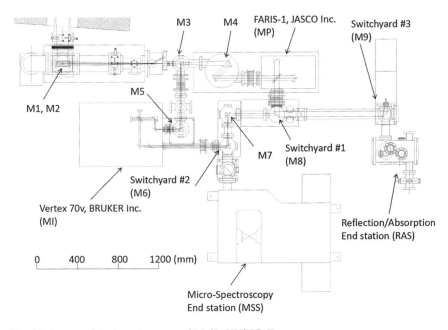

Fig. 4.8 Layout of the beam transport of BL6B, UVSOR-II

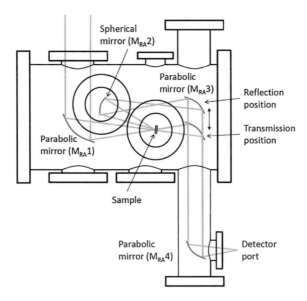

Fig. 4.9 Top view of the optical path of the reflection/absorption end-station (RAS). The reflection/transmission experimental setup can be changed by moving the parabolic mirror $M_{RA}3$

end-station (RAS) by a plane mirror (M8) in the switchyard #1. The light after the MI is guided to the microspectroscopy end-station (MSS) and the RAS by a plane mirror (M6) in the switchyard #2.

4.4.1 THz/IR Reflection/Absorption Spectroscopy Using SR

Figure 4.9 shows the optics of the RAS. Using the optics, both a reflection spectroscopy and transmission spectroscopy are available. The two experiments can be changed by adding a parabolic mirror $M_{RA}3$ without breaking the vacuum. Samples and all optical components are located in a high vacuum chamber ($\leq 10^{-6}$ Pa) evacuated by a turbomolecular pump (300 l/s). There is no optical window in front of the samples. This has the advantages of no limitation of the photon energy and no interference by optical windows. An evaporator for gold film on the sample surface is added for measurement of absolute values. An automatic measurement system is also added to measure the relative intensity between the sample and a reference mirror at another sample space [11]. The spectral distribution of IR/THz SR using RAS in the wavenumber region below 1000 cm^{-1} is shown in Fig. 4.10. The spectrum is covered down to ~4 cm^{-1} (~0.13 THz) by using three kinds of combinations of light source (UVSOR-II), FTIR (FARIS-1 or Vertex 70v), and detector (InSb hot-electron bolometer or Si bolometer). The typical intensity in this region is about 100 times higher than thermal sources.

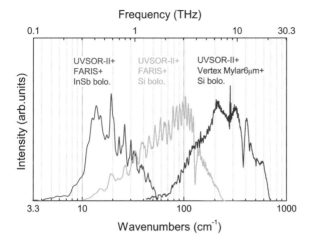

Fig. 4.10 Spectral distribution of the combination of light source (UVSOR-II), FTIR (FARIS-1 or Vertex 70v), and detector (InSb hot-electron bolometer or Si bolometer) in the THz region of BL6B, UVSOR-II. The lowest accessible wavenumber is about 4 cm^{-1} (~0.13 THz)

4.4.2 THz/IR Microspectroscopy Using SR

In the visible region, optical lenses are generally added for focusing. However, optical lenses cannot be used in the THz/IR region, because astigmatism appears: different refractive indexes occur because if the presence of impurities and many absorption lines from the lattice and molecular vibration modes. In the case of THz/IR, one mirror system consisting of one concave and one convex spherical mirror, namely Schwarzschild mirrors, is used (Fig. 4.11d). This system is used because it eliminates the astigmatism and gives high reflectivity in the THz/IR region due to Drude reflection of metals.

IR/THz SR is a powerful tool for microspectroscopy and imaging not only in the IR region but also in the THz region because of its high brilliance. Almost all of the world's IR/THz SR beamlines provide commercial IR microscopes that are easy to use. However, because the Schwarzschild mirrors in commercial IR microscopes are small (typically 2 inches in diameter), the microscopes are available only in the mid-IR region, i.e., they are not suitable for the THz region. However, we have installed a new microscope in order to cover the IR region and down to the THz region (Fig. 4.11) [12]. Some of the main features are: (1) a large working distance, because experiments with specific requirements (at very low temperatures, under high pressures, for near-field spectroscopy, etc.) need to be performed; and (2) availability of the THz region, because the quasiparticle states of correlated materials and finger print vibration modes of proteins appear only in this spectral range. In order to achieve these goals, we use one set of large Schwarzschild mirrors (diameter = 140 mm, numerical aperture = 0.5, working distance = 106 mm, magnification = ×8) to reduce the diffraction effect in the THz region.

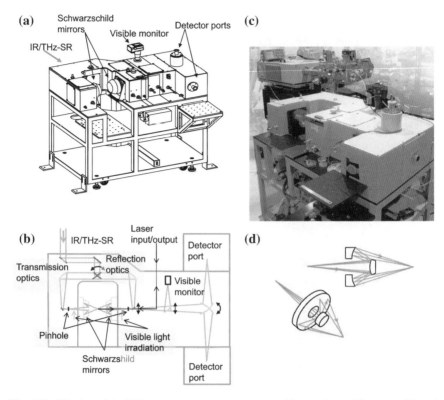

Fig. 4.11 3D view of the THz microspectroscopy apparatus (**a**); top view of the optics (**b**), and photo (**c**) of the THz/IR microspectroscopy end-station (MSS) of the IR/THz beamline (BL6B) at UVSOR-II. **d** indicates the configuration of a Schwarzschild mirror

After installing the THz microscope, we checked the spatial resolution in the different wavenumber ranges covered by the different FTIR beam splitters (Fig. 4.12). The spatial resolution was recorded by using a Bruker Vertex 70v interferometer and the transmission configuration. In the wavenumber region above $500\,cm^{-1}$, the globar lamp gives good contrast. However, the use of UVSOR-II increases the intensity by 10^2 and decreases the beam size by about 10 times. This implies that the brilliance of the IR/THz SR is 10^3 times higher than that of thermal sources. This is consistent with the brilliance measured at the first focusing point P1 (Fig. 4.5b). If a pinhole 500 μm in diameter is installed at a counter focus position of the Schwarzschild mirror, the beam size becomes the diffraction limit. In the wavenumber region below $600\,cm^{-1}$, in which a Mylar 6 μm beam splitter is used, a good beam profile was observed with UVSOR-II, in contrast with the very weak peak given by a globar lamp. In addition, use of a Mylar 23 μm beam splitter covering the lower wavenumber region below $200\,cm^{-1}$ gave good contrast and intensity in the peak image. The cut-off wavenumber for UVSOR-II is $40\,cm^{-1}$ ($150\,cm^{-1}$ in the case of the SPring-8 IR beamline [13]). This indicates that the THz microscope using UVSOR-II is a

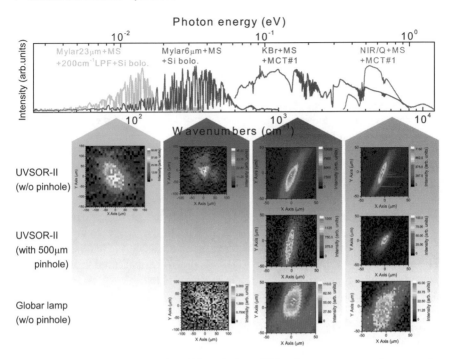

Fig. 4.12 Spectral distribution and spatial resolution of THz microscope at several wavenumber regions using UVSOR-II compared with a globar lamp. Labels in the *top* figure indicate the combinations of beam splitter of the FTIR (Bruker), low-pass filter, and detector. Labels on the *left side* indicate the light sources and apertures at other focuses of the Schwarzschild mirror. Fine structures in the spectral distributions are due to the absorption of water vapor and can be reduced by nitrogen purge. See the text for details

very efficient tool for microspectroscopy in both the IR and the THz region. This instrument is used for electrodynamics studies of **strongly correlated materials** at low temperatures and under high pressures [6], and also for molecular vibration imaging of living cells.

4.5 Application of THz Spectroscopy Using SR: Temperature-Dependent THz Spectra of SmB$_6$

SmB$_6$, which is a typical **Kondo semiconductor**, has been studied over three decades. In the previous studies, two different energy gap sizes were proposed one about 5 meV and the other about 15 meV [14]. However, a recent study using higher purity samples revealed that the lower energy absorption band originates from impurities, because the lower energy absorption becomes small with increasing purity [15, 16]. The higher energy gap is concluded to be intrinsic and to originate from hybridization between the localized Sm $4f$ state and the Sm $5d$ conduction band, so-called c–f **hybridization** [17]. In the case of a static c–f hybridization, a **rigid band model** suitably explains the physical properties. However, the specific heat curve cannot be explained by a rigid band model [18]. This means that there are temperature-dependent parameters in the electronic structure as well as in the charge dynamics. Angle-resolved photoemission spectra also indicate complex temperature dependence of electronic structure [19].

On the other hand, the magnetic excitation at 14 meV increases with decreasing temperature below 20 K, as observed by using inelastic neutron scattering [20]. The magnetic excitation energy is similar to the energy gap observed in optical spectra. Therefore, if the c–f hybridization bands exist at the energy gap edge, the carriers should have the same temperature dependence. Optical reflection spectroscopy is good as a probe for investigating the character of carriers. In the case of SmB_6, the signal from the thermally excited carriers appears in the THz region. The change in the fitting parameters of the Drude function as a function of temperature must reflect the property at the energy gap edge. We, therefore, measured the temperature dependence of the reflectivity spectrum in the THz region.

The thick lines in Figure 4.13a show the temperature dependence of the reflectivity spectrum [$R(\omega)$] of SmB_6. At $T = 5$ K, the reflectivity does not approach unity with decreasing photon energy. This shows an insulating character. With increasing temperature, the reflectivity at the lower energy portion increases. The high reflectivity originates from the Drude component due to the presence of thermally excited

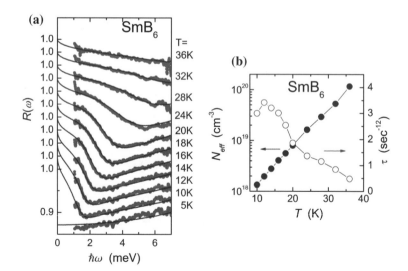

Fig. 4.13 **a** Temperature dependence of reflectivity spectrum of SmB_6 in the THz region (*thick lines*) and fitting functions by the combination of a Drude and two Lorentz functions (*thin lines*). Successive curves are offset by 0.03 for clarity. **b** Parameters obtained for the effective electron number (N_{eff}) and the relaxation time (τ) of the Drude function as a function of temperature

carriers. The Drude component indicates the electronic structure of the thermally excited area near the energy gap edge.

The curves fitted by using one Drude and two Lorentz functions are indicated by the thin solid lines in Fig. 4.13a. The fitting function is as follows:

$$\hat{\varepsilon}(\omega) = \varepsilon_\infty - \frac{4\pi N_{\mathrm{eff}} e^2}{m(\omega^2 - i\omega/\tau)} + \sum_{i=1}^{n} \frac{4\pi N_i e^2}{m\{(\omega_{i0}^2 - \omega^2) - i\omega/\tau_i\}}$$

$$R(\omega) = \left| \frac{1 - \hat{\varepsilon}(\omega)^{1/2}}{1 + \hat{\varepsilon}(\omega)^{1/2}} \right|^2$$

Here, $\hat{\varepsilon}(\omega)$, $\hat{n}(\omega)$, and $R(\omega)$ are a complex dielectric function, a complex refractive index and a reflectivity spectrum, respectively. ε_∞ is the sum of ε_1 above the measured energy region, N_{eff} and τ are the effective number and the relaxation time of the carriers, respectively, and m the rest mass of an electron. N_i, ω_{i0}, and τ_i are the intensity, resonance frequency, and relaxation time, respectively, of ith bound state, including energy gaps in the electronic structure and optical phonons. The two Lorentz functions were set to the main gap at 15 meV and to the impurity state at 5 meV. The fitting functions reproduced the experimental curves well. The obtained N_{eff} and τ, derived from the fitting parameters, are shown as functions of temperature (Fig. 4.13b). If an activation-type behavior ($N_{\mathrm{eff}} \propto \exp(-2\Delta/k_{\mathrm{B}}T)$, where Δ is the energy gap and k_{B} is the Boltzmann constant) is expected, then log N_{eff} should be proportional to $1/T$. However, log N_{eff} is roughly proportional to T, as shown in Fig. 4.13b. This indicates that the energy gap cannot be explained by the rigid band model, but it may also indicate that the energy gap shrinks with increasing temperature. A two-state model could be assumed to explain this behavior [21]. On the other hand, τ rapidly increases on cooling below 20 K. This temperature is coincident with that of the growth of the magnetic excitation at 14 meV [20]. This means that thermal excitation of the carriers is strongly related to magnetic excitation at 14 meV.

4.6 Applications of THz Microspectroscopy Using SR

4.6.1 Experimental Procedure of THz Spectroscopy Under High Pressure

THz spectroscopy under high pressures has been performed by using the MSS, as shown in Fig. 4.14. To detect the charge dynamics, the optical reflection spectroscopy [$R(\omega)$] in the THz region must be performed, as mentioned in the previous section. High pressure is produced by using a diamond anvil cellDAC. The experimental setup for reflection spectroscopy at high pressure and low temperature at MSS is

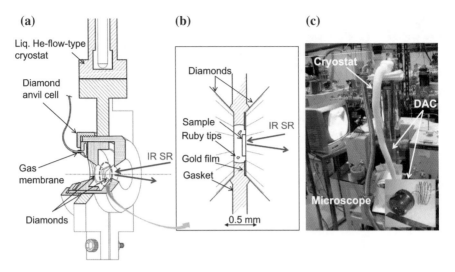

Fig. 4.14 Schematic figure of a diamond anvil cell (DAC) with a gas membrane combined with a liquid-helium-flow-type cryostat for reflection spectroscopy at high pressure and low temperature (**a**); the sample setup in the DAC (**b**); and photo of the experimental apparatus and the DAC (**c**)

shown in Fig. 4.14. Because the sample area in the pressure cell is smaller than 1 mm in diameter, we have to use a microscopic technique. THz spectroscopy under high pressures is difficult to perform using conventional FTIRs.

A membrane-type DAC (Diacell® OptDAC-LT, easyLab Technology Ltd.) was employed to produce high pressures on the samples. Figure 4.14c is a photo of the setup combined with the THz microscope. Pressure was applied to samples at low temperatures by using the He gas inlet to the membrane. Because the culet plane areas of diamond and of the membrane are 1.13 and 1030 mm^2, respectively, the pressure on samples is about 10^3 times larger than the helium gas pressure. The maximum pressure at the sample position is about 8 GPa. A sample with a typical size of $0.4 \times 0.4 \times 0.05$ mm^3 was set in a DAC with Apiezon-N grease as a pressure medium, with gold film as a reference and ruby tips for a pressure reference. The pressure was calibrated by ruby fluorescence measurement.

4.6.2 Electronic Structure of SmS Under High Pressure

SmS is an insulator (the so-called "black phase") with a gap size of 1000 K ($\sim$80 meV) at ambient pressure [22]. Above about 0.7 GPa, the sample color changes to gold (golden phase) and the Sm-ion changes from divalence to mixed valence [23]. To investigate the mechanism of this transition, we performed THz reflection spectroscopy under pressure.

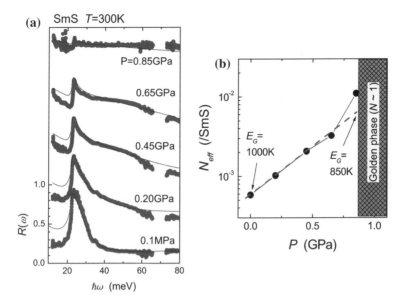

Fig. 4.15 **a** Pressure dependence of the reflectivity spectrum [$R(\omega)$] of SmS (*thick lines*) in the black phase at 300 K. The fitting curves from the combination of a Drude and the Lorentz functions are plotted as *thin solid lines*. Successive curves are offset by 0.5 for clarity. **b** Pressure dependence of energy gap evaluated by Drude and Lorentz fitting of $R(\omega)$ spectra. See the text for details

The $R(\omega)$ obtained for SmS at 300 K as a function of pressure is shown by the thick lines in Fig. 4.15a. At ambient pressure, the spectrum indicates an insulating one, because the low energy limit does not approach unity and a clear large peak due to the TO-phonon between the Sm^{2+} and S^{2-} ions appears. When pressure was applied, the background intensity increased with increasing pressure. The background indicates the appearance of carriers. Therefore, the carrier density increased with increasing pressure. The fitting curves of the combination of a Drude function and Lorentz functions are shown by the thin solid lines in the same figure. The Lorentz function was set to reproduce the TO-phonon. To fit the spectra obtained, only the N_{eff} in the Drude function is changed; all of the other parameters (τ in the Drude function and all parameters in the Lorentz function) are fixed. The N_{eff} obtained is plotted in Fig. 4.15b. The log N_{eff} is proportional to the pressure up to 0.65 GPa. This pressure dependence indicates that the energy gap closes with increasing pressure. The energy gap size (2Δ) at ambient pressure was evaluated to be about 1,000 K [22]. By evaluating the gap size from N_{eff}, the energy gap at the black-golden phase boundary is 850 K; the gap then suddenly closes to 100 K in the golden phase by first-order transition [24].

4.7 Conclusion and Outlook

In this chapter, we have pointed out the high brilliance of the infrared (IR) and terahertz (THz) synchrotron radiation (SR) and have introduced the characteristic optics by which "three-dimensional magic mirror (3D-MM)" is used to focus the bending-magnet radiation. The 3D-MM is an ideal optical component for focusing bending-magnet radiation, because not only can the synchrotron light be focused down to a diameter close to the electron beam size but also the time structure can be reduced to that of the electron beam length. Using 3D-MM, some advanced optical experiments have so far been performed (e.g., highly precise THz reflection spectroscopy, and THz spectroscopy at high pressure and low temperature). 3D-MM can also be used for time-resolved spectroscopy because of the conservation of the time structure of electron beams.

Recently, THz coherent SR (THz-CSR) has been developed in storage rings. THz-CSR, which is emitted when the electron beam length or the longitudinal structure is shorter than the wave length, has intensity several orders of magnitude higher than that of normal SR [25]. The light has properties not only of high intensity but also full coherence. The light can be used for new experiments (e.g., THz-pump probe spectroscopy, and THz phase contrast imaging) [26]. Such new experiments will open up a new frontier in spectroscopy.

Acknowledgments We would like to thank UVSOR staff members for their devoted support and Dr. Takuya Iizuka for providing Fig. 14. High purity samples of SmB_6 and SmS were provided by Prof. Satoru Kunii and Prof. Noriaki K. Sato, respectively.

References

1. T. Nanba, Y. Urashima, M. Ikezawa, M. Watanabe, E. Nakamura, K. Fukui, H. Inokuchi, Int. J. Infrared Millimeter Waves **7**, 759 (1986)
2. M. Katoh, M. Hosaka, A. Mochihashi, J. Yamazami, K. Hayashi, Y. Hori, T. Honda, K. Haga, Y. Takashima, T. Koseki, S. Koda, H. Kitamura, T. Hara, T. Tanaka, AIP Conf. Proc. **705**, 49 (2004)
3. S. Kimura, E. Nakamura, T. Nishi, Y. Sakurai, K. Hayashi, J. Yamazaki, M. Katoh, Infrared Phys. Tech. **49**, 147 (2006)
4. S. Kimura, E. Nakamura, J. Yamazaki, M. Katoh, T. Nishi, H. Okamura, M. Matsunami, L. Chen, T. Nanba, AIP Conf. Proc. **705**, 416 (2004)
5. H. Kimura, T. Moriwaki, N. Takahashi, H. Aoyagi, T. Matsushita, Y. Ishizawa, M. Masaki, S. Ohishi, H. Okuma, T. Nanba, M. Sakurai, S. Kimura, H. Okamura, N. Nakagawa, T. Takahashi, K. Fukui, K. Shinoda, Y. Kondo, T. Sata, M. Okuno, M. Matsunami, R. Koyanagi, Y. Yoshimatsu, Nucl. Instrum. Meth. A **467–468**, 441 (2001)
6. S. Kimura, H. Okamura, J. Phys. Soc. Jpn. **82**, 021004 (2013)
7. T. Yaji, Y. Yamamoto, T. Ohta, S. Kimura, Infrared Phys. Tech. **51**, 397 (2008)
8. J.D. Jackson, *Classical Electrodynamics*, 3rd edn. (John Wiley & Sons, New York, 1999)
9. R. Lopez-Delgado, H. Szwarc, Opt. Commun. **19**, 286 (1976)
10. S. Kimura, H. Kimura, T. Takahashi, K. Fukui, Y. Kondo, Y. Yoshimatsu, T. Moriwaki, T. Nanba, T. Ishikawa, Nucl. Instrum. Meth. A **467–468**, 437 (2001)

11. S. Kimura, JASCO Report **50**, 6 (2008). (in Japanese)
12. S. Kimura, Y. Sakurai, E. Nakamura, T. Mizuno, AIP Conf. Proc. **879**, 595 (2006)
13. Y. Ikemoto, T. Moriwaki, T. Hirono, S. Kimura, K. Shinoda, M. Matsunami, N. Nagai, T. Nanba, K. Kobayashi, H. Kimura, Infrared Phys. Tech. **45**, 369 (2004)
14. G. Travaglini, P. Wachter, Phys. Rev. B **29**, 893 (1984)
15. T. Nanba, H. Ohta, M. Motokawa, S. Kimura, S. Kunii, T. Kasuya, Physica B **186–188**, 440 (1993)
16. S. Kimura, T. Nanba, S. Kunii, T. Kasuya, Phys. Rev. B **50**, 1406 (1994)
17. G. Aeppli, Z. Fisk, Comments Condens. Matter Phys. **16**, 155 (1992)
18. T. Kasuya, K. Takegahara, Y. Aoki, K. Hanzawa, M. Kasaya, S. Kunii, T. Fujita, N. Sato, H. Kimura, T. Komatsubara, T. Furuno, J. Rossat-Mignod, Valence Fluctuation in Solids, p. 215, (North-Holland, Amsterdam, 1981).
19. H. Miyazaki, T. Hajiri, T. Ito, S. Kunii, S. Kimura, Phys. Rev. B **86**, 075105 (2012)
20. P.A. Alekseev, J.M. Mignot, J. Rossat-Mignod, V.N. Lazukov, I.P. Sadikov, E.S. Konovalova, YuB Paderno. J. Phys. Condens. Matter **7**, 289 (1995)
21. B. Gorchunov, N. Sluchanko, A. Volkov, M. Dressel, G. Knebel, A. Loidl, S. Kunii, Phys. Rev. B **59**, 1808 (1999)
22. K. Matsubayashi, K. Imura, H.S. Suzuki, T. Mizuno, S. Kimura, T. Nishioka, K. Kodama, N.K. Sato, J. Phys. Soc. Jpn. **76**, 064601 (2007)
23. J.L. Kirk, K. Vedam, V. Narayanamurti, A. Jayaraman, E. Bucher, Phys. Rev. B **6**, 3023 (1972)
24. T. Mizuno, S. Kimura, K. Matsubayashi, K. Imura, H.S. Suzuki, N.K. Sato, J. Phys. Soc. Jpn. **77**, 113704 (2008)
25. S. Kimura, E. Nakamura, K. Imura, M. Hosaka, T. Takahashi, M. Katoh, J. Phys.: Conf. Ser. **359**, 012009 (2012)
26. S. Kimura, E. Nakamura, M. Hosaka, T. Takahashi, M. Katoh, AIP Conf. Proc. **1234**, 63 (2010)

Part II
Ultrafast and Coherent Measurement

Chapter 5
Time-Resolved X-Ray Diffraction Studies of Coherent Lattice Dynamics Using Synchrotron Radiation

Yoshihito Tanaka

5.1 Introduction

When a surface of materials is irradiated by a femtosecond pulsed laser, the electronic state is excited, and then coherent lattice vibrational mode in collective atomic motion appears during the energy relaxation to thermal energy. In semiconductors, coherent lattice vibrational modes arise in non-thermal and hot-excitation regime [1]. The speed of the collective atomic motion is dependent on the mode as well as material. The frequency of optical phonon mode generally lies above terahertz region; acoustic mode has relatively low and wide-range frequency. The fast phonon modes have intensively been observed by the photoreflectivity or transmissivity measurement with a femtosecond pulsed laser through the change in macroscopic relative permittivity [2]. On the other hand, X-ray diffraction has a powerful method to observe such atomic motion in materials directly. An intense short-pulse X-ray beam is required to observe such fast atomic motion in materials.

Synchrotron radiation (SR), which is generated from a relativistic electron beam accelerated by applying electric and magnetic fields, is characterized by its well-collimated beam and wide range tunability of photon energy from THz radiation to X-rays [3]. In the past decades, significant progress has been seen in recent development of SR sources, which can provide an X-ray beam with high intensity, high spatial coherence, and ultrafast pulsed structure with various repetition rate [4, 5]. The intense short-pulse X-rays with high repetition rate have opened the opportunity to conduct time-resolved X-ray diffraction measurement of the fast atomic motion [6, 7].

Time-resolved measurement techniques are mainly categorized into two types. Illumination of a sample with a pulsed X-ray beam enables to take a snapshot of X-ray diffraction pattern, referring to a stroboscopic method. By changing the time

Y. Tanaka (✉)
RIKEN SPring-8 Center, 1-1-1 Kouto Sayo-cho, Sayo-gun, Hyogo 679-5148, Japan
e-mail: yotanaka@riken.jp

K. Shudo et al. (eds.), *Frontiers in Optical Methods*,
Springer Series in Optical Sciences 180, DOI: 10.1007/978-3-642-40594-5_5,
© Springer-Verlag Berlin Heidelberg 2014

interval between a laser irradiation and an X-ray snapshot, atomic arrangement in progress during reaction can be observed with frame step. Alternatively, a beam from highly repetitive pulsed photon source can be regarded as a continuous wave (CW) or quasi-CW, which is useful for measurement of X-ray diffraction intensity changes with detectors. These methods are selected for required time-resolution in order to observe a series of atomic motions or lattice dynamics such as coherent phonons as collectively vibrating motion, and lattice deformation including expansion and shrinkage.

Investigations on lattice dynamics using SR time-resolved X-ray diffraction have actively been reported in recent decades. The optical phonons [8] and acoustic phonons [9] produced in a single crystal induced by a femotosecond pulsed laser irradiation, and picoseconds melting in semiconductor single crystals [10, 11] have been reported. We also conducted the measurement at SPring-8 SR facility, and succeeded in observation of acoustic pulses generation [12] and the echoes [13]. A nanosecond response in the piezo-electric devices induced by electric field has also been investigated using SR [14].

In this chapter, coherent lattice dynamics induced by femtosecond pulsed laser irradiation and the time-resolved observation with X-ray diffraction using SR are described. First, the present status of SR facility and the characteristics are reviewed in Sect. 5.2, and then measurement techniques including time-resolved measurement technique for various time scales, and X-ray diffraction method for the investigation of lattice dynamics are described in Sect. 5.3. In Sect. 5.4, the experimental components and the techniques are explained through the development at SPring-8 facility. Section 5.5 shows the examples on the observation of an acoustic pulse, optical and acoustic phonons, and is finally followed by the future perspectives in Sect. 5.6.

5.2 Synchrotron Radiation

5.2.1 Present Status of SR Facilities

In the first-generation (1960s), the SR which wreaks particle beam energy loss and should have been reduced in a particle accelerator, had been, in contrast, used as a light source for materials science at a high-energy particle accelerator facility. In 1970s, the accelerators for SR were constructed. SOR-RING and PF (photon factory) are typical facilities in Japan, and are categorized in the second generation. In 1990s, the synchrotron facility whose main radiation source is an undulator (a device to wiggle the electrons in the accelerator) is constructed as the third generation facilities, including SPring-8 (Super Photon Ring-8 GeV, Harima in Hyogo, see Fig. 5.1), APS (Advanced Photon Source, USA), and ESRF (European Synchrotron Radiation Facility, France). In this chapter, the research at the third-generation SR facility is mainly described. In 2000s, compact third-generation sources appeared: Although the ring size and electron energy are relatively small, hard X-ray generation is possible with an undulator having shorter period and strong magnetic field.

Fig. 5.1 Aerial photograph of the SPring-8 campus. An 8 GeV storage ring (SPring-8, The circumference is 1,436 m) and an X-ray free electron laser facility (SACLA, The length is about 700 m) are located in the same campus in Hyogo, Japan. ©RIKEN/JASRI

They are, in other words, ecological SR facilities. On the other hand, the plan of the development aiming at super machines has also been promoted: They are an energy recovery linac (ERL) and the fourth-generation, an X-ray free electron laser (XFEL), both of them are composed of a linear accelerator, producing femtosecond X-ray pulses with small size and divergence. As for the XFEL development, the construction with single pass amplification scheme without a resonator, which is called SASE (Self Amplified Spontaneous Emission) has been completed and the research phase has just started at LCLS (Stanford, USA) [15] and SACLA (SPring-8, Hyogo, Japan, see Fig. 5.1) [16]. The construction is also in progress at European XFEL facility (DESY, Germany) [17, 18].

5.2.2 Characteristic Features of SR

The characteristic features of SR produced at third-generation storage ring and linear accelerator (linac) facilities are briefly reviewed in this section.

Undulator radiation generated at SPirng-8, which is one of typical third-generation SR facility, is described as a typical example. The SR is characterized by variable wavelengths, collimated beam, polarizability, and pulsed time structure. The wavelengths are tuned by changing the undulator gap, and are picked up with a monochromator with narrow bandwidth of about $\Delta E/E = 10^{-4}$ in the range of 5–37 keV at a hard X-ray undulator beamline. An X-ray beam with several tens of keV is also available with the harmonic radiation from an undulator. The beam divergence is about 10 µrad, which gives the size of 0.5 mm at about 50 m downstream from

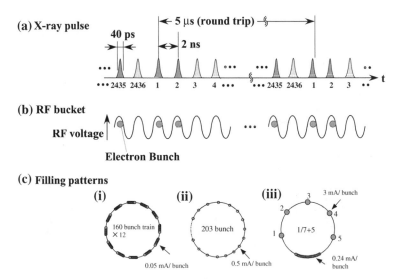

Fig. 5.2 Time structure of (**a**) X-ray SR pulses and (**b**) the RF buckets in the SPring-8 storage ring. (**c**) The examples of the filling patterns operated in the ring

the source. The high degree of polarization can be controlled with the variety of undulator type to linear and circular polarizations, and also with an optics of phase retarder to give a fast change of circular polarization between right and left directions.

The time structure is characterized by a picosecond pulse duration and a high repetition rate of MHz-GHz. Figure 5.2a shows the time-structure of SR in the SPring-8 storage ring. The pulse duration is about 40 ps in full width at half maximum (FWHM), which is slightly dependent on the bunch current. The pulse intervals are determined by the RF frequency of 508.58 MHz of the cavity for acceleration (Fig. 5.2b) of electron bunches in the storage ring. The intervals are thus between 2 ns and 5 μs. The filling patterns, as shown in Fig. 5.2c, are operated on request.

The X-ray intensity from a 27 m-long undulator is, for example, 10^{14} photons for one second, corresponding to the power of 200 mW. The peak power is about 600 W for the X-ray pulse with a bunch current of 3 mA, pulse duration of 50 ps, and the photon energy resolution of 10^{-4}.

For linac-based sources, the smaller beam size, smaller divergence, and shorter pulse duration are achievable. The XFEL located at SPring-8 campus (SACLA) is expected to produce an intense beam with a pulse duration of 100 fs $\sim$ 5 fs, and the repetition rate of 60 Hz.

5.3 Time-Resolved X-Ray Diffraction

Time-resolved X-ray diffraction is achieved by combination of "Time-resolved measurement method" and "X-ray diffraction", as the name just indicates. This section thus describes the two techniques separately: time-resolved measurement technique using a SR beam (in Sect. 5.3.1), and the X-ray diffraction method mainly applied for the investigation of lattice dynamics (in Sect. 5.3.2).

5.3.1 Time-Resolved Measurement Method

As SR has the time-structures as shown in Sect. 5.2.2, there are several techniques for the time-resolved measurement methods dependent on requested time resolution and the scales. Figure 5.3 shows the time-resolution and the corresponding measurement techniques.

For wide range observation with time-resolution of $>$ 1ns, the time-resolution of detectors are utilized for a quasi-continuous wave (CW) source, because of the high repetition frequency (MHz to GHz) (see Fig. 5.3a). In Fig. 5.3, the typical method is described where pulse counting signals using an avalanche photodiode (APD) are processed with MCS (multichannel scaler). The time-resolution is about 1 ns, which is determined by the rise time of the APD.

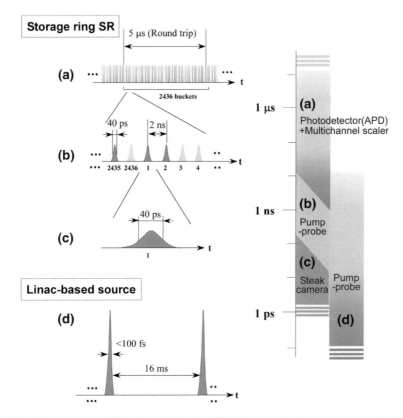

Fig. 5.3 Time-structures of SR pulses (*left*) and available time-resolved measurement method with the time resolution (*right*)

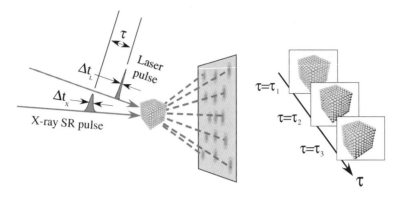

Fig. 5.4 A schematic illustration of time-resolved X-ray diffraction with laser-pump and SR probe method

For time range of 40 ps to nanoseconds, pump-probe method is available using the pulsed structure of the SR, where the snapshot of X-ray scattering patterns is taken at an appropriate time after an impulse stimulus (see Fig. 5.3b). Figure 5.4 schematically shows the laser-pump SR-probe method. The time-resolution is determined by the stimulus pulse width of Δt_L and an X-ray probe pulse width of Δt_X.

Accumulation is required when the signal intensity is weak, the high precision of timing synchronization between the stimulus and the SR pulse is required. In the SPring-8 storage ring, the time-resolution is about 40 ps (FWHM), which is determined by Δt_X.

For faster time-scale where the SR pulse can be recognized to be CW (<40 ps) as shown in Fig. 5.3c, an X-ray streak camera [19, 20] is useful for time-evolution measurement of the phenomena completed within 40 ps. The time-resolution is in the range of a few ps to several hundreds of fs, which is dependent on the photon energy.

For femtosecond time resolution, linac-based femotosecond pulsed X-ray sources will be available, when the timing between pulsed sources is controlled or measured with femotosecond time precision in the pump-probe measurement system (Fig. 5.3d).

5.3.2 Observation of Lattice Dynamics with X-Ray Diffraction

For investigation of structural dynamics with time-resolved X-ray diffraction method, an ideal way is to determine the crystal structure at each time by the structure analysis with single-crystal X-ray diffraction. However, the way spends too much time to obtain clear data. Thus, precise measurement of change in the intensity and its distribution of a particular Bragg spot has practically been conducted to investigate optical and acoustic phonon oscillations and lattice deformation. The explanation is given below with schematic examples.

Figure 5.5a, a' represent a crystal lattice in the reciprocal space and real space, respectively. In order to obtain an X-ray diffraction intensity distribution as shown in Fig. 5.5a, the space in the reciprocal space should be covered with the composition of vectors of incident and diffracted X-ray beams by changing the angles between them. Generally, in the single crystal X-ray diffraction, diffraction intensity distribution of a lot of Bragg spots is obtained by changing sample angle and diffracted angle with respect to the direction of an incident monochromatic X-ray beam. Another diffraction measurement methods are powder diffraction where a monochromatic X-ray beam irradiates the many sample particles with various orientations, and white Laue crystallography where a single crystal is exposed with a white beam having wide-band spectrum.

For the investigation of crystal lattice dynamics, diffraction patterns as shown in Fig. 5.5a are obtained for each time during lattice change sometimes with small amount of change. So, a change in diffraction distribution of one Bragg spot (for example, 002 diffraction spot in Fig. 5.5) is precisely measured, assuming that, the lattice is homogeneously deformed, or other Bragg spots are also regularly changed in a same manner.

"How the lattice deformation, acoustic phonon, and optical phonon appear on the diffraction pattern is described below".

To know lattice deformation, the peak position shift of the Bragg diffraction spot is measured in a reciprocal map. Consider the case that an initial crystal lattice shown in Fig. 5.5a or a' expands in the direction of (001) as shown in Fig. 5.5b'. When the Bragg spot with an index of 002 is observed, the peak position moves in the direction where the scattering vector becomes shorter on the reciprocal map shown in Fig. 5.5b.

As for acoustic phonon, it appears at the position indicated by vectorial sum of the reciprocal lattice vector and the momentum vector of acoustic phonon in Fig. 5.5c. The acoustic phonon vibration corresponding to the momentum is observed as temporal behavior.

In optical phonon, atomic displacement occurs alternatively. Due to the motion, a period of the atomic arrangement is changed to be between doubled and single periods. As a result, an X-ray diffraction spot in Fig. 5.5 blinks at a position where the diffraction intensity is originally reduced or enhanced according to the extinction rule. Figure 5.5d shows the case that a structure factor with an index of 001 is reduced by extinction rule. Generation of optical phonon induces temporal change in the amount of crystal structure factor, leading the diffraction intensity of 001 vibrates with the phonon frequency.

The atomic motions described above are also characterized by the time-scales. The response time of lattice deformation is typically around nanoseconds. The periods of acoustic and optical phonons typically lie around picosecond and femtosecond time-scale, respectively. So the suitable time-resolved measurement method should be selected from Sect. 5.2.1. The typical examples will be shown in Sect. 5.5.

92   Y. Tanaka

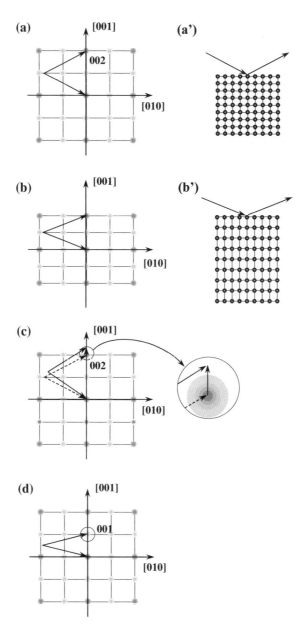

Fig. 5.5 X-ray diffraction scheme applied for the investigation on lattice dynamics. The reciprocal map (**a**) for initial lattice structure (**a'**), is modified to (**b**) by the lattice expansion in the [001] direction. Acoustic phonon behavior is detected by scanning around the Bragg spot as shown in (**c**). The optical phonon in which the atoms alternately make displacement is detected by monitoring of the diffraction spot related to the extinction rule as shown in (**d**). This illustration represents a crystal of zinc blende structure as an example

5.4 Equipment Components and the Techniques

This section shows the components of the apparatus and the required technique for time-resolved X-ray diffraction method, to take the case of SPring-8 as an example.

5.4.1 Equipment Components

A schematic diagram of experimental setup for time-resolved X-ray diffraction measurement in an SR facility of SPring-8 is shown in Fig. 5.6 [7]. It is composed of a short pulse laser giving pulsed stimulus to a sample, an X-ray diffractometer, and a timing control system including data acquisition system. A photograph in the upper right of Fig. 5.6 shows a femtosecond pulsed laser system in SPring-8, including a mode-locked Ti:sapphire laser oscillator, and a regenerative amplifier. As shown in Fig. 5.3a, in the case that the SR is considered to be quasi-CW light, the diffraction intensity is integrated as the timing when a sample is irradiated by the laser pulse is determined to be zero. In the case of pump-probe and streak method, as shown in Fig. 5.3b–d, synchronization between a pulsed laser and a SR source is indispensable. In Sect. 5.4.2, the timing control techniques including synchronization are described.

5.4.2 Timing Control Technique Between Laser and SR Pulses

This section describes the technique of timing control of a femtosecond pulsed laser and storage ring SR X-ray pulse [21, 22]. The timing control consists of the timing synchronization (phase synchronization), time delay control, and repetition rate control. The required precision of timing control is determined to be better than the SR pulse duration of $\sim$40 ps. Each technical components are described bellow.

5.4.2.1 Phase Synchronization Technique

As shown in Fig. 5.6, a mode-locked pulsed laser is synchronized to the RF master oscillator controlling an RF cavity accelerating electron bunches in the storage ring, in order to achieve the laser-SR synchronization. As the repetition rate of a mode-locked laser is determined by the cavity length, a position of the cavity mirrors is controlled with a piezoelectric translator by a feedback circuit to lock the phase between the RF trigger signal and the laser pulse train. In SPring-8, a mode-locked Ti:sapphire laser oscillator with pulse duration of 80 fs is phase-locked to a trigger signal with a frequency (84.76 MHz) of 1/6 of an RF master oscillator (508.58 MHz) of the storage ring. The timing precision is better than 5 ps (see Fig. 5.7(i)).

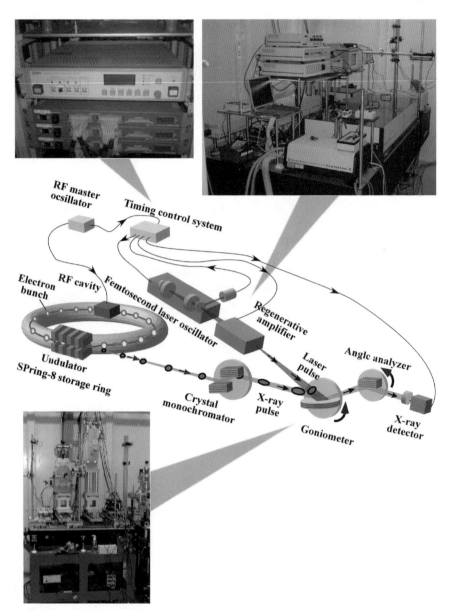

Fig. 5.6 A schematic illustration of time-resolved X-ray diffraction measurement system at the SPring-8 SR facility. The system is mainly composed of an SR source, a femtosecond pulsed laser, the timing control system, and an X-ray diffractometer

The repetition rate of regenerative amplifier is determined to be ∼1 kHz, which is figured out by the RF master frequency being divided by the multiple number

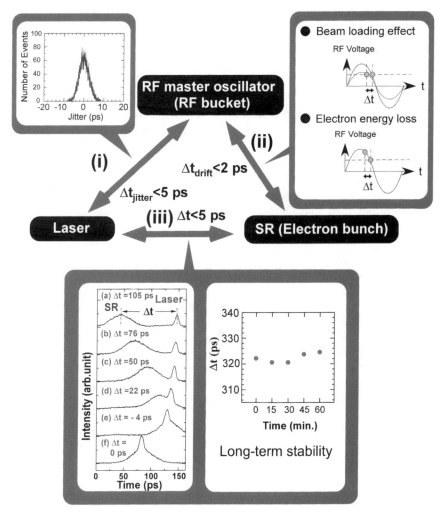

Fig. 5.7 Timing precision achieved at the SPring-8 storage ring. The synchronization precision was evaluated to be better than 5 ps

of bucket (see Fig. 5.2b) of 2436. The laser pulse can thus meet the X-ray pulse generated from a particular bunch in the storage ring.

The behavior of the phase between the RF cavity voltage and the electron bunch should also be considered, in addition to the synchronization between the laser pulse and RF master oscillator. In the storage ring operation, the accelerator has a feedback system for electrons to turn stably round and round with energy loss by the SR generation. Due to the feedback system, the phase drift occurs between the RF cavity voltage and electron bunch timing. The phase drift is mainly originated from the following two issues: (i) transient drop of RF voltage by beam loading, and (ii) the variation of electron energy loss by undulator radiation power change (Fig. 5.7(ii)). In fact, when the undulator gap of 14 segments was fully open and closed to change the SR output power in the SPring-8 storage ring, the timing drift by $\sim$100 ps was observed [22]. As for the beam loading effect, the position of electron bunches in the RF bucket (see Fig. 5.2b) is stable with respect to each bunch, as the shift is dependent on the bunch address. Thus, locking a laser shot at the X-ray pulses from a particular electron bunch makes us to avoid the influence of the beam loading effect. The undulator power dependence is practically out of issue, because it is impossible to make several undulator gaps fully opened and closed simultaneously in usual user beamtime. However, for precise measurement, a feedback control system should be requested to compensate the timing drift.

Figure 5.7(iii) shows the result of the evaluation of timing precision by using an X-ray streak camera. The time drift was less than 5 ps over hours, which is efficiently shorter than the SR pulse duration of about 40 ps.

5.4.2.2 Time Delay Control Technique

In laser-pump SR-probe measurement, the time interval, τ, between laser and SR pulses is changed as shown in Fig. 5.4. Generally, optical delay is applied to a laser beam, to control the time interval. However, as the place of optical delay component is sometimes far from the sample position in an SR facility, larger optical delay may make a miss-alignment at the sample. Then, in the SPring-8 facility, an electronic circuit delay system for RF trigger was developed by changing the RF phase continuously [23]. This method enables to make an arbitrary delay with a few picoseond precision [24].

5.4.2.3 Repetition Rate Control Technique

For pump-probe experiments, the repetition rates of laser and SR pulses for sample irradiation should be the same. The laser system for excitation of the sample, is usually equipped with an amplifier to gain higher power as shown in Sect. 5.4.2.1. Accordingly, the repetition rate of the amplified pulses is reduced to $\sim$1 kHz frequency. Furthermore, the interval should be longer than the recovery time of the target phenomenon. Thus, a MHz repetition rate of SR as shown in Sect. 5.2.2, also should be reduced to $\sim$1 kHz. An X-ray rotating chopper is sometimes used to control the repetition rate of SR. In the rotating chopper, as a thick blade is required to stop the X-ray beam due to high transmittance, the rotation axis is perpendicular to the X-ray beam direction (see the example of Fig. 5.8(iii)). Another way to pick up X-ray pulses is to apply an electronic gate to the signal from a fast X-ray detector such as an avalanche photodiode (APD), which can observe X-ray pulsed structure (see Fig. 5.8). By selecting the filling pattern of electron bunches (see Fig. 5.2c(iii)) to obtain larger interval, a 2D camera with electronic gate of sub-microsecond duration may also be available.

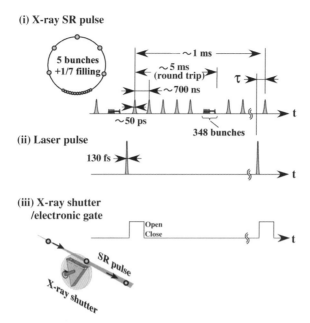

Fig. 5.8 Time chart of X-ray SR, the pulsed laser, and the gate of an X-ray shutter (or electronic gate) for a pump-probe measurement with a time delay of τ

5.5 Examples of Observation of Lattice Deformation and Phonons

This section describes time-resolved X-ray diffraction measurements using SR for three types of coherent atomic motions: Acoustic pulse generation, acoustic phonon and optical phonon.

5.5.1 Acoustic Pulses and the Echoes

Figure 5.9a shows the example to observe an acoustic pulse and the echo generated in a Si(111) single crystal wafer, when the surface is irradiated by a laser with a pulse duration of 130 fs and a wavelength of 800 nm [13]. In this case, data acquisition system is not synchronized with the SR pulses, and the X-ray signal from an APD detector is accumulated using an MCS (see Sect. 5.3.1). The time-resolution is about 1 ns. X-ray diffractometers determine the diffraction angle, 2θ, and the sample angle, ω. The angle scan with keeping the relation of $2\theta = 2\omega$, catches the lattice expansion and shrinkage in the direction of reciprocal lattice vector (in this case, it coincides with the surface normal) [12, 25]. The relation among Bragg angle, θ, the peak shift,

Fig. 5.9 Acoustic pulse
echoes observed by the Bragg
peak shift of Si 333 (**a**) and
GaAs 004 (**b–d**). The time-
resolved diffraction data were
obtained with MCS (**a** and
b), and pump-probe method
(**c** and **d**)

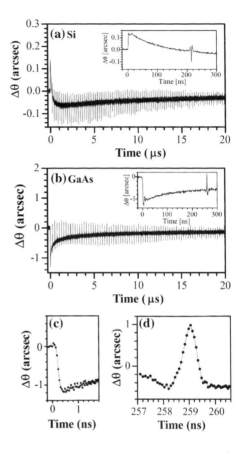

$\Delta\theta$, and the rate of crystal lattice spacing, $\Delta d/d$, is given by,

$$\Delta d/d = -\Delta\theta/\tan\theta. \qquad (5.1)$$

The time-evolution of the peak shift, $\Delta\theta$, of Si 333 Bragg diffraction is shown
in Fig. 5.9a. From 5.1, the plus and minus direction of $\Delta\theta$ correspond to shrinkage and expansion, respectively. The dimension of $\Delta\theta$ is arcsec, where 1 arcsec =
1/3600°. Just after laser irradiation, the lattice is compressed, and the impulse excitation induces acoustic pulse. The relaxation to thermal energy makes the lattice
space expand around 100 ns after the laser irradiation. The interval of around 200 ns
corresponds to the period when the acoustic pulse generated at the front surface of a
1 mm-thick wafer propagates and is reflected at the rear surface and then go back to
the front surface.

Figure 5.9b shows the experimental result for a 625 μm-thick GaAs(001) wafer.
The lattice expands by the laser irradiation, and the deformation of the generated
acoustic pulse has an opposite phase with respect to the pulses seen in a Si wafer.

The acoustic pulse duration is about $1 \sim 2$ ns, whose pulse shape cannot be well analyzed due to the time-resolution of 1 ns with an MCS-APD method. Then, pump-probe method with a time-resolution of 40 ps is applied for initial strain and the first echo pulse, as shown in Fig. 5.9c, d. The precise measurement of the pulse shape reveals that the wave packet undergoes broadening and deformation with the propagation and the diffraction effect. The analysis of echo pulse shape makes it possible to know the roughness of the rear surface, just like a sonar exploring objects in the sea [13].

The precise observation of propagation time of the acoustic pulse also enables the measurement of the speed of sound in solid materials. The example is shown for GaAs single crystal [24]. High resolution and high precision measurement of the propagation time can be achieved by a pump-probe method and the clock delay circuit which has a high precision even for long-time delay (Sect. 5.4.2). Figure 5.10 shows the result of the arrival time measurement of the pulse echoes. This experiment has a total system precision of 20 ps, leading to the precision of 4–5 orders of magnitude. High precision measurement of the thickness of the wafer can thus determine the speed of sound in the material with high precision.

5.5.2 Acoustic Phonons

Careful observation of the acoustic pulse as described in Sec. 5.5.1 with high time-resolution and high-momentum resolution makes it possible to observe coherent acoustic phonon mode [9, 12]. In this section, observation of acoustic phonon for a single crystal of indium antimonide (InSb) is described, which has been reported by the group of APS [9]. As shown in Fig. 5.5c, temporal behavior of acoustic phonon appears at the position on the reciprocal space, which is indicated by the vectorial sum of the wavevector of acoustic phonon and reciprocal lattice vector, **G**. The relation between the deviation, $\Delta\theta$, from Bragg angle, θ and acoustic phonon frequency of Ω propagating parallel to **G** is given by,

Fig. 5.10 Time evolution of X-ray diffraction intensity for acoustic pulse echoes of GaAs wafer. Arrival time of third acoustic pulse echo is evaluated to be 806.97 ± 0.01 ns

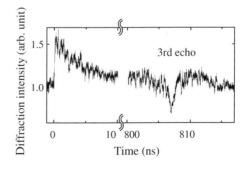

Fig. 5.11 Acoustic phonons
in InSb, observed through the
time dependence of diffraction
intensity at crystal angles of
0, 20, and 40 arcsec from
the Bragg peak. The arrows
indicate the oscillation (Data
is taken from [9])

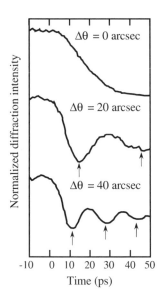

$$\Omega = v \mid \mathbf{G} \mid \Delta\theta / \tan\theta, \qquad (5.2)$$

where v is a speed of sound in the crystal. The acoustic phonon with higher vibration frequency is observed at larger $\Delta\theta$ in the X-ray diffraction observation. Figure 5.11 shows the time-evolution of X-ray diffraction at the off-set angle, $\Delta\theta$, from the Bragg angle, θ, which is observed with an X-ray streak camera. This method corresponds to (c) in Fig. 5.3. It is found in Fig. 5.11 that for $\Delta\theta = 0$, no vibration appears; for larger $\Delta\theta$, vibration frequency becomes higher, which is consistent with 5.2.

5.5.3 Optical Phonons

In order to observe optical phonons in time domain, sub-picosecond time resolution is required due to the fast atomic motion [8]. The X-ray pulse duration obtained in the third-generation synchrotron radiation facility is ∼1 ps at most. A linac-based SR source is required to obtain a sub-picosecond X-ray SR pulse. The experiment at the test facility for development of SASE-FEL (see Sect. 5.2.1) in SLAC (Stanford) is described in this section. In the test facility, femtosecond pulsed electron beam is guided to an undulator to obtain X-ray pulse with a duration of 170 fs (FWHM). As described in Sect. 5.4.2.1, in pump-probe experiment, timing synchronization is required to have a precision of better than their pulse duration. However, sub-picosecond synchronization is rather difficult. Another scheme is applied: the shot-by-shot timing between the pump laser and the electron bunch is recorded, and the data point is rearranged according to the recorded timing after completing the

Fig. 5.12 Optical phonon oscillation (as indicated with arrows) observed in the diffraction intensity of Bi 111 (Data is taken from [8])

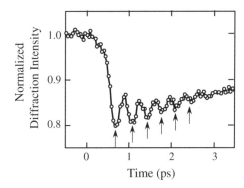

measurement. Here, the shot-by-shot timing measurement was achieved by electro-optical effect [26]. Figure 5.12 shows the experimental result of X-ray diffraction intensity oscillation due to optical phonon vibration in a bismuth crystal. The optical phonon induced by a laser irradiation causes fast change in 111-Bragg diffraction intensity with a period of around 350 fs. In a bismuth crystal, the atom located around the center of the unit cell is slightly shifted from the center position. So the 111-diffraction is not perfectly inhibited, and the coherent alternative motion of atoms induces modulation of structure factor, resulting in the diffraction intensity modulation. The paper reported that softening of the optical phonon is induced by the intense laser irradiation.

5.6 Summary and Perspectives

In this chapter, the time-resolved X-ray diffraction using a SR pulsed source and the research on a coherent lattice dynamics with the SR source are described. Development of SR sources is still in progress as shown in Sect. 5.2.1, expecting a shortly pulsed and more intense light source. Other short-pulse X-ray sources such as laser Compton sources, laser plasma sources, and X-ray diodes [6] are also candidates for time-resolved diffraction experiments. In fact, a laser-plasma X-ray source has been used for time-resolved X-ray diffraction experiment on coherent optical phonon [27, 28]. For soft X-ray region, higher harmonics of ultrashort pulsed laser is useful. Combination of laser and SR source is also effective to obtain short pulses: For example, a laser slicing method, in which short and intense laser pulse is introduced into a storage ring, gives the energy modulation in a part of electron bunch to generate short-pulse SR in the particular direction [29, 30], and has succeeded in observation of optical phonon oscillation [31]. Laser seeding with harmonic generation of SASE-FEL [32] is also one of challenges to obtain a stable and shorter pulse X-ray beam in the near future.

Development of the time-resolved X-ray diffraction method with high
time-resolution and high precision will open a door to investigate collective atomic
motions with comparing the atomic arrangement and the oscillation in permittivity due to coherent phonons [2, 33, 34]. Furthermore, the combination of X-ray
microbeam (or focusing) technique and time-resolved technique [35], may make it
possible to observe the propagation of lattice distortion along the surface. The investigation of lattice dynamics is expected to be in progress along with the development
of SR experimental techniques.

References

1. J. Shah, *Ultrafast Spectroscopy of Semiconductors and Semiconductor Nanostructures* (Springer, New York, 1998)
2. M. Hase, K. Ishioka, J. Demsar, K. Ushida, M. Kitajima, Phys. Rev. B **71**, 184301 (2005)
3. H. Wiedemann, *Synchrotron Radiation* (Springer, Heidelberg, 2003)
4. J.A. Nielsen, *Elements of Modern X-ray Physics* (John Wiley & Sons, West Sussex, 2001)
5. D.M. Mills, *Third-Generation Hard X-Ray Synchrotron Radiation Sources: Source Properties, Optics, and Experimental Techniques* (John Wiley & Son, New York, 2002)
6. J.R. Helliwell, R.M. Rentzepis Time-Resolved Diffraction (Oxford Scientific Publications, Oxford, 1997).
7. Y. Tanaka, T. Ishikawa, in Femtosecond Beam Science, ed. by M. Uesaka. Time-Resolved Diffraction Using SR (Imperial College Press/World Scientific Pub Co Inc, NJ, 2005), p. 318.
8. D.M. Fritz, D.A. Reis, B. Adams, R.A. Akre, J. Arthur, C. Blome, P.H. Bucksbaum, A.L. Cavalieri, S. Engemann, S. Fahy, R.W. Falcone, P.H. Fuoss, K.J. Gaffney, M.J. George, J. Hajdu, M.P. Hertlein, P.B. Hillyard, M. Hoegen, M. Kammler, J. Kaspar, R. Kienberger, P. Krejcik, S.H. Lee, A.M. Lindenberg, D. Meyer, T. Montagne, E.D. Murray, A.J. Nelson, M. Nicoul, R. Pahl, J. Rudati, H. Schlarb, D.P. Siddons, K.S. Tinten, Th Tschentscher, D. Linde, J.B. Hastings. Science **315**, 633 (2007)
9. A.M. Lindenberg, I. Kang, S.L. Johnson, T. Missalla, P.A. Heimann, Z. Chang, J. Larsson, P.H. Buckbaum, H.C. Kapteyn, H.A. Padmore, R.W. Lee, J.S. Wark, R.W. Falcone, Phys. Rev. Lett. **84**, 111 (2000)
10. K.S. Tinten, C. Blome, C. Dietrich, A. Tarasevitch, M. Horn von Hoegen, D. Linde, A. Cavalleri, J. Squier, M. Kammler, Phys. Rev. Lett. **87**, 225701 (2001)
11. A. Rousse, C. Rischel, S. Fourmaux, I. Uschmann, S. Sebban, G. Grillon, Ph Balcou, E. Foerster, J.P. Geindre, P. Audebert, J.C. Gauthier, D. Hulin. Nature **410**, 65 (2001)
12. Y. Tanaka, Y. Uozaki, K. Nozaki, K. Ito, K. Yamasaki, H. Terauchi, I. Takahashi, K. Tahara, T. Ishikawa, J. Phys. Conf. Ser. **287**, 012018 (2011)
13. Y. Hayashi, Y. Tanaka, T. Kirimura, N. Tsukuda, E. Kuramoto, T. Ishikawa, Phys. Rev. Lett. **96**, 115505 (2006)
14. A. Grigoriev, D.-H. Do, D. Kim, C.-B. Eom, B. Adams, E. Dufresne, P. Evans, Phys. Rev. Lett. **96**, 187601 (2006)
15. L.F. DiMauro, J. Arthur, N. Berrah, J. Bozek, J.N. Galayda, J. Hastings, J. Phys. Conf. Ser. **88**, 012058 (2007)
16. T. Shintake, H. Tanaka, T. Hara, T. Tanaka, K. Togawa, M. Yabashi, Y. Otake, Y. Asano, T. Bizen, T. Fukui, S. Goto, A. Higashiya, T. Hirono, N. Hosoda, T. Inagaki, S. Inoue, M. Ishii, Y. Kim, H. Kimura, M. Kitamura, T. Kobayashi, H. Maesaka, T. Masuda, S. Matsui, T. Matsushita, X. Marechal, M. Nagasono, H. Ohashi, T. Ohata, T. Ohshima, K. Onoe, K. Shirasawa, T. Takagi, S. Takahashi, M. Takeuchi, K. Tamasaku, R. Tanaka, Y. Tanaka, T. Tanikawa, T. Togashi, S. Wu, A. Yamashita, K. Yanagida, C. Zhang, H. Kitamura, T. Ishikawa, Nat. Photonics **2**, 555 (2008)

17. J. Andruszkow et al., Phys. Rev. Lett. **85**, 3825 (2000)
18. V. Ayvazyan et al., Eur. Phys. J. D **37**, 297 (2006)
19. Y. Tanaka, S. Adachi, Detectors for timing measurements II -Streak cameras and fast photodiodes. J. Jpn. Soc. Synchrotron Radiat. Res. [in Japanese] 22, 77 (2009).
20. T. Hara, Y. Tanaka, H. Kitamura, T. Ishikawa, Rev. Sci. Instrum. **71**, 3624 (2000)
21. Y. Tanaka, T. Hara, H. Kitamura, T. Ishikawa, Rev. Laser Eng. [in Japanese] 30, 525 (2002).
22. Y. Tanaka, T. Hara, H. Kitamura, T. Ishikawa, Rev. Sci. Instrum. **71**, 1268 (2000)
23. Y. Tanaka, T. Ohshima, Y. Fukuyama, N. Yasuda, J. Kim, H. Osawa, S. Kimura, T. Togashi, T. Hara, H. Kamioka, Y. Moritomo, H. Tanaka, M. Takata, H. Sengoku, E. Nonoshita, AIP Conf. Proc. **1234**, 951 (2010)
24. Y. Fukuyama, N. Yasuda, J. Kim, H. Murayama, T. Ohshima, Y. Tanaka, S. Kimura, H. Kamioka, Y. Moritomo, K. Toriumi, H. Tanaka, K. Kato, T. Ishikawa, M. Takata, Rev. Sci. Instrum. **79**, 045107 (2008)
25. Y. Hayashi, N. Tsukuda, E. Kuramoto, Y. Tanaka, T. Ishikawa, J. Synchrotron Radiat. **12**, 685 (2005)
26. A. Cavalieri, D. Fritz, S. Lee, P. Bucksbaum, D. Reis, J. Rudati, D. Mills, P. Fuoss, G. Stephenson, C. Kao, D. Siddons, D. Lowney, A. MacPhee, D. Weinstein, R. Falcone, R. Pahl, J.A. Nielsen, C. Blome, S. Dusterer, R. Ischebeck, H. Schlarb, H. S.-Schrepping, Th. Tschentscher, J. Schneider, O. Hignette, F. Sette, K. S. -Tinten, H. Chapman, R. Lee, T. Hansen, O. Synnergren, J. Larsson, S. Techert, J. Sheppard, J. Wark, M. Bergh, C. Caleman, G. Huldt, D. Van der Spoel, N. Timneanu, J. Hajdu, R.A. Akre, E. Bong, P. Emma, P. Krejcik, J. Arthur, S. Brennan, K. Gaffney, A. Lindenberg, K. Luening, J. Hastings. Phys. Rev. Lett. **94**, 114801 (2005)
27. K.S. Tinten, C. Blome, J. Blums, A. Cavalleri, C. Dietrich, A. Tarasevitch, I. Uschmann, E. Foerster, M. Kammler, M. Hoegen, D. Linde, Nature **422**, 287 (2003)
28. K. Nakamura, S. Ishii, S. Ishitsu, M. Shiokawa, H. Takahashi, K. Dharmalingam, J. Irisawa, Y. Hironaka, K. Ishioka, M. Kitajima, Appl. Phys. Lett. **93**, 061905 (2008)
29. R.W. Schoenlein, S. Chattopadhyay, H.H.W. Chong, T.E. Glover, P.A. Heimann, C.V. Shank, A.A. Zholents, M.S. Zolotorev, Science **287**, 2237 (2000)
30. P. Beaud, P. Beaud, S.L. Johnson, A. Streun, R. Abela, D. Abramsohn, D. Grolimund, F. Krasniqi, T. Schmidt, V. Schlott, G. Ingold, Phys. Rev. Lett. **99**, 174801 (2007)
31. S.L. Johnson, P. Beaud, C.J. Milne, F.S. Krasniqi, E.S. Zijlstra, M.E. Garcia, M. Kaiser, D. Grolimund, R. Abela, G. Ingold, Phys. Rev. Lett. **100**, 155501 (2008)
32. G. Lambert, T. Hara, D. Garzella, T. Tanikawa, M. Labat, B. Carre, H. Kitamura, T. Shintake, M. Bougeard, S. Inoue, Y. Tanaka, P. Salieres, H. Merdji, O. Chubar, O. Gobert, K. Tahara, M.-E. Couprie, Nature Phys. **4**, 296 (2008)
33. M. Hase, K. Mizoqichi, H. Harima, S. Nakashima, M. Tani, K. Sakai, M. Hangyo, Appl. Phys. Lett. **69**, 2474 (1996)
34. D.H. Hurley, R. Lewis, O.B. Wright, O. Matsuda, Appl. Phys. Lett. **93**, 113101 (2008)
35. Y. Tanaka, Y. Fukuyama, N. Yasuda, J. Kim, H. Murayama, S. Kohara, H. Osawa, T. Nakagawa, S. Kimura, K. Kato, F. Yoshida, H. Kamioka, Y. Moritomo, T. Matsunaga, R. Kojima, N. Yamada, K. Toriumi, T. Ohshima, H. Tanaka, M. Takata, Jpn. J. Appl. Phys. 48, 03A001 (2009).

Chapter 6
Coherent Phonon Dynamics in Carbon Nanotubes

Keiko Kato, Katsuya Oguri and Masahiro Kitajima

6.1 Introduction

Moore's law, which states that the number of transistors on a chip will double approximately every two years, has driven technology and industry for decades. For Moore's law to remain valid there must be innovations as regards the scaling down of devices. However, further scaling down has faced serious limitations related to fabrication techniques as the device dimension reaches nano-order, which is comparable to the carrier scattering length. To overcome these limitations, the further reductions in device scale have been attempted by turning to alternative materials. Prime candidates for this purpose are low-dimensional carbon materials such as **carbon nanotubes** and graphene, which have attracted a lot of attention due to their exceptional electronic properties including quasi-ballistic transport and a current-carrying capability [1–4].

Carbon nanotubes (CNTs), which consist of rolled-up sheets of graphene, exhibit unique electronic, mechanical, and optical properties. Depending on the helical arrangement of the graphene sheet (i.e., **chirality**), single-walled carbon nanotubes (SWCNTs) can be either metallic or semiconducting [5]. If the band gap in SWCNTs can be controlled, SWCNTs would be promising candidates for future nano-scale

K. Kato (✉) · K. Oguri
NTT Basic Research Laboratories, Nippon Telegraph and Telephone Corporation,
3-1 Morinosato Wakamiya, Atsugi, Kanagawa 243-0198, Japan
e-mail: kato.keiko@lab.ntt.co.jp

K. Oguri
e-mail: oguri.katsuya@lab.ntt.co.jp

M. Kitajima
R&D Group, LxRay Co. Ltd., Kosien-2bancho, Nishinomiya 633-8172, Japan
e-mail: kitajima@LxRay.jp

M. Kitajima
Faculty of Engineering, Yokohama National University, Tokiwadai 79-5,
Hodogaya, Yokohama 240-8501, Japan

K. Shudo et al. (eds.), *Frontiers in Optical Methods*,
Springer Series in Optical Sciences 180, DOI: 10.1007/978-3-642-40594-5_6,
© Springer-Verlag Berlin Heidelberg 2014

transistor devices [6]. Moreover, metallic SWCNTs exhibit the long elastic mean free paths (typically on the order of micrometers) [1, 2], which is limited by electron backscattering with optical or zone-boundary phonons. Thus, the electron-phonon interaction is the fundamental bottleneck for ballistic transport in SWCNTs.

Another important property of CNTs is the ultrafast photo-induced response. After the generation of excited-carriers by irradiation of light, carrier scatterings such as electron-electron, electron-phonon, or carrier-impurity scatterings occur within a few hundred femtoseconds [7]. The fast optical response makes CNTs a good candidate for the realization of, for example, ultrafast switching devices [8, 9] and ultrashort pulse lasers [10–12]. In terms of applications, it is important to understand carrier and phonon dynamics in CNTs. Pump-probe techniques with ultrashort laser pulses are powerful ways to obtain information about carrier and phonon dynamics in CNTs. In particular, coherent phonons, which are a large number of phonons in one mode with a constant phase relation as the result of impulsive excitation by ultrashort laser pulses [13], could reveal us the transient dynamics and develop further potentials of CNTs. By observing coherent phonons, we can obtain dynamical information, such as the phonon generation and decay dynamics [14], the carrier-phonon interaction [15], and the initial phase of the phonon oscillation [16, 17]. By interfering coherent phonons, Kim et al. succeeded in exciting coherent phonons of the SWCNTs with a specific chirality [16, 18]. Coherent phonon provides transient carrier and phonon dynamics, which cannot be obtained with conventional continuous-wave (cw) Raman measurements, and also presents the way to control physical properties of SWCNTs.

In contrast to conventional optical measurements with cw lasers, the measurement of coherent phonons requires us to modify the detection scheme because of the small signal, which appears within several pico seconds. Besides, we need to consider the sample separation because the broad energy band width of the ultrashort laser pulse excites SWCNTs with different chiralities simultaneously and makes it difficult to obtain the detailed information from SWCNTs with a specific chirality. In this review, we describe time-resolved studies of ultrafast coherent phonon dynamics in SWCNTs. The chapter is organized as follows. In Sect. 6.2, we review the fundamental physical properties of SWCNTs to understand optical responses of SWCNTs. In Sect. 6.3, we introduce the generation and detection of coherent phonons and describe our experimental setup. In Sect. 6.4, we present a time-resolved investigation of aligned SWCNTs with sub-10-fs laser pulses to study the polarization dependence of coherent phonons in SWCNTs. In Sect. 6.5, we report the observation of coherent phonons in metallic SWCNTs, which are separated from mixed samples and investigate ultrafast dynamics of free carriers and phonons.

6.2 Carbon Nanotube

There are many excellent reviews that deal with the basic physical properties of carbon nanotubes [19, 20]. This section provides a brief overview of some of the fundamental properties of carbon nanotubes.

6.2.1 Electronic Structure

A single-walled carbon nanotube (SWCNT), which consists of a rolled-up graphene sheet, can be either metallic or semiconducting depending on its helical arrangement (i.e., chirality) [19]. The electronic structure of SWCNTs is obtained by using periodic boundary conditions on the electronic structure of graphene. Figure 6.1 shows the electronic density of states for (a) metallic and (b) semiconducting SWCNTs [21]. Depending on the chirality, the density of state has a non-zero (zero) value for metallic (semiconducting) SWCNTs near the Fermi level, located at $E = 0$.

Of particular interest is the fact that the singularities appear in the density of states due to the one-dimensional structure. Because of the singularities, the electronic density of states of SWCNTs takes maxima, which appear as a near mirror image set of spikes around the Fermi level, as shown in Fig. 6.1. Transitions occur with irradiation of light between the mirror image spikes in the valence and conduction bands [23]. The optical absorption spectra of SWCNTs exhibit peaks due to the discreteness in the electronic density of states [22]. The absorbance strongly depends on the light polarization. The parallel polarization of light with respect to the SWCNT axis enhances absorption (i.e., the **antenna effect**) [24], whereas perpendicular polarization suppresses it (i.e., the **depolarization effect**) [25]. The transition energy

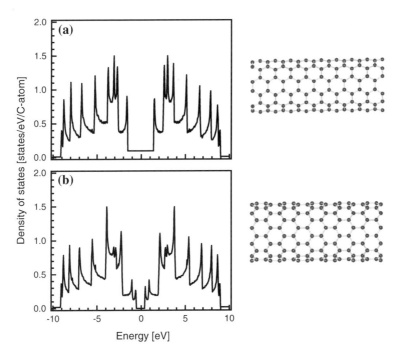

Fig. 6.1 Electronic density of states for **a** metallic and **b** semiconducting SWCNTs. Data points are taken from [21]. The right figures show corresponding SWCNT structures

Fig. 6.2 The gap energy values are plotted for possible configuration as a function of the diameter (Kataura plot) (Reproduced from [22] with permission from Elsevier)

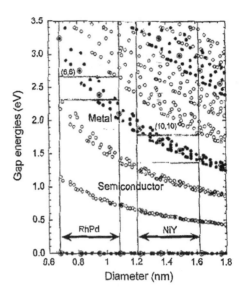

(i.e., the gap energy between the mirror spikes) depends on the chirality. Kataura et al. plot transition energy as a function of SWCNT diameter as shown in Fig. 6.2 [22], which is called as **Kataura plot**. In the Kataura plot, we see that metallic and semiconducting SWCNTs with the same diameter have different transition energies. Therefore, we can selectively excite each of them by changing the incident light wavelength.

6.2.2 Phonon

The phonon dispersion relations of SWCNTs are obtained theoretically with the zone folding of those for graphene [19, 20, 26]. Some phonon modes, which have zero frequency at the Γ point (i.e., the phonon vector, $q = 0$) in graphene, have a non-zero frequency and are Raman active in SWCNTs due to zone folding. The calculated atomic displacements, frequencies, and symmetries for the (10, 10) SWCNT in [20] are shown in Fig. 6.3a–d.

The phonon properties of SWCNTs have been extensively investigated with Raman spectroscopy [27]. Raman spectra are obtained as a function of the energy difference between the incident and scattered light. When the photon energy of the incident light is close to the gap energy between the singularities, the Raman intensity is enhanced by the **resonance Raman effect** [22]. The resonance Raman effect can be induced by the irradiation of visible light, because the transition energy of SWCNTs is in the visible region as shown in the Kataura plot (Fig. 6.2). From the Raman spectra, we can obtain information on the frequency and lifetime of phonons

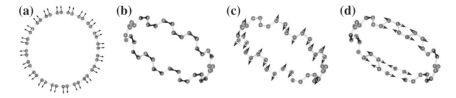

Fig. 6.3 The calculated Raman mode atomic displacements, frequencies, and symmetries for the (10, 10) nanotube modes. The frequencies are taken from [20]. **a** A_{1g} 165 cm^{-1}, **b** A_{1g} 1587 cm^{-1}, **c** E_{1g} 1585 cm^{-1} and **d** E_{2g} 1591 cm^{-1}

in SWCNTs. The Raman spectra of SWCNTs generally have several peaks as shown in Fig. 6.4. The observed peaks correspond to (i) the radial breathing mode (RBM) (Fig. 6.3a), (ii) the D mode, and (iii) the G mode. In the following section, we describe each mode in detail.

Radial Breathing Mode (RBM)

In the RBM, which has A symmetry, all the carbon atoms move in phase in the radial direction as if the tube were breathing (Fig. 6.3a). The observation of the RBM constitutes the identification of SWCNTs in the sample because the RBM becomes a Raman active mode as a result of the zone folding. The RBM exhibits a peak around 100–300 cm^{-1} in Raman spectra (inset in Fig. 6.4). The RBM, whose frequency is inversely proportional to the SWCNT diameter, is used to estimate the diameter [28].

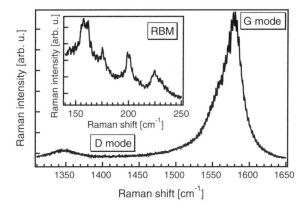

Fig. 6.4 Raman spectrum of SWCNTs. The inset shows the RBM spectrum

G Mode

The G mode corresponds to C-C stretching modes (Fig. 6.3b–d) and exhibits several peaks around $1585 \, cm^{-1}$ (Fig. 6.4) [26, 29, 30]. The split structure of the G mode in SWCNTs originates from the phonon dispersion of graphite [26, 31, 32]. In graphite, the longitudinal optical (LO) and transverse optical (TO) phonon modes are degenerate at the Γ point and exhibit a single Lorentzian peak in the Raman spectrum. In SWCNTs, the degeneracy is broken by the curvature effect [31]. Consequently, the G mode of the SWCNTs splits into lower (G^-) and higher (G^+) frequency modes [31, 33]. In semiconducting SWCNTs, the $G^+(G^-)$ mode corresponds to the TO(LO) phonon mode. This reversal occurs in metallic SWCNTs due to the softening of the LO phonon mode through the interaction with free electrons near the Fermi level (Kohn anomaly) [34]. The interaction between the LO phonon mode and electrons has an asymmetric line shape in the G^- mode, which is known as the Breit-Wigner-Fano (BWF) shape [35, 36]. Furthermore, for each of the G^- and G^+ modes, the periodic boundary condition in the circumferential direction of a SWCNT defines the angular momentum of vibrational motion along the nanotube axis of 0, 1, and 2, corresponding to the A, E_1, and E_2 symmetry phonon modes, respectively (Fig. 6.3b–d) [26, 31]. We mention the detection of phonons with different symmetries later.

The split structure of the G mode can be used for the characterization of SWCNTs. The frequency of the G^- mode decreases with increases in the diameter, whereas that of the G^+ mode is independent of the diameter [32, 33]. Thus, the splitting of the G^+ and G^- modes is inversely proportional to the square of the SWCNT diameter due to the curvature effect. In addition, the G^- mode in metallic SWCNTs has a lower frequency than that in semiconducting SWCNTs with the same diameter [33]. From the difference between the frequencies of the G^+ and G^- modes, we can roughly estimate the diameter and identify SWCNTs as metallic or semiconducting.

D Mode

Defects induce an additional Raman peak around $1350 \, cm^{-1}$ (Fig. 6.4), which is called the D mode [37]. The D mode, which has non-zero phonon wave vector ($q \neq 0$), cannot be observed in a perfect crystal, but becomes Raman active through scattering with defects [38, 39]. The relative intensity between the G and D modes, which becomes high with an increase in the number of defects, is used to estimate the defect density [37].

Resonance Raman Effect

Resonance Raman effects, which enhance the intensity of Raman spectra, occur when the photon energy of either incident or scattered light is close to the transition energy of the SWCNTs. Figure 6.5 shows schematic diagrams of resonance Raman processes

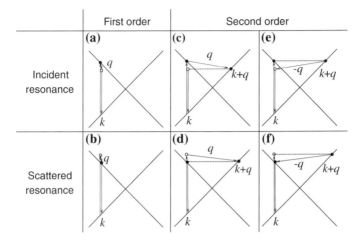

Fig. 6.5 Schematic diagram for the first order resonant Raman (**a** and **b**) and double resonant Raman processes (**c–f**) proposed by Saito in [40]. The upper and lower figures show the incident light resonance and scattered light resonance, respectively. The *solid dots* are resonance energy states

[39, 40]. The crossed lines correspond to the electronic energy dispersion around the K point in graphite. With the irradiation of visible light, the transition occurs around the K point (the crossing point in Fig. 6.5), which corresponds to the Fermi level. The black dots are resonant states where the energy gap between the valence and conduction bands is the same as the photon energy of the incident or scattered lights. For example, in Fig. 6.5a, an electron with momentum k is resonantly excited with the incident laser photon, whose energy matches the transition energy from the valence band to the conduction band. Then, the electron and hole recombine accompanied by the emission of the phonon. In Fig. 6.5b, the resonance occurs with scattered light, whose photon energy matches the transition energy from the conduction band to the valence band. In the first-order Raman process (Fig. 6.5a, b), where the scattering between the light and the phonon occurs only once, there is no momentum transfer because the light emission occurs through the recombination of an electron and hole. Therefore, the first-order Raman process produces the phonon only at Γ point (i.e., the wave vector $q = 0$) because the momentum of the light is much larger than that of the phonon. The RBM and G-mode phonons, which have non-zero frequency at the Γ point, are generated by the first-order Raman process.

The D mode, on the other hand, cannot be produced by the first-order Raman process because it has a non-zero wave vector. Saito et al., explained the **double resonance Raman process** contributes to the generation of the D mode as shown in Fig. 6.5c–e [39]. In the double resonance Raman process, (i) an electron with momentum k is excited with the incident light, (ii) the electron is then scattered to a state with momentum $k + q$ with a phonon (or defect), and finally (iii) the electron is scattered back to the state with momentum k with a defect (or phonon) to recombine with a hole. One of the two scattering processes ($k \rightarrow k+q, k+q \rightarrow k$) is inelastic scattering that accompanies the phonon emission process (a solid line with an energy change), and the other is elastic scattering process mediated with the defect (a dotted line without an energy change). Consequently, the Raman shift corresponds to the one-phonon energy. The intensity of the D mode is comparable to that of the RBM or G mode because the scattering occurs through the resonant states during the double resonance Raman process.

Polarized Raman Spectroscopy

The symmetry of phonons can be identified with **polarized Raman spectroscopy**. To discuss the phonon symmetry in SWCNTs, we describe the Raman intensity in terms of the phonon symmetry.

The observed Raman intensity (I_R) is given by

$$I_R \propto \left| \mathbf{e_s} \cdot \underline{\alpha} \cdot \mathbf{e_L} \right| \tag{6.1}$$

where $\mathbf{e_s}$, $\mathbf{e_L}$, and $\underline{\alpha}$ are the polarizations of the incident light, and the scattered light, and the Raman tensor. In (6.1), the phonon symmetry, which determines the point group of the Raman tensor, dominates the Raman selection rule. For example, for totally symmetric phonon modes, (i.e., A symmetry), only diagonal matrix elements in the Raman tensor have non-zero values whereas off-diagonal matrix elements have non-zero values for non-totally symmetric phonon modes (ex. E_1 or E_2 symmetries). Thus, we can selectively detect a specific phonon mode by choosing the polarization of the incident or scattered light with respect to the SWCNT axis (i.e., polarized Raman spectroscopy).

In fact, Jorio et al. succeeded in detecting E_1- and E_2-symmetry G modes from an isolated SWCNT with a polarized Raman measurement [29, 30]. The intensity of these non-totally symmetric modes is much lower than that of A symmetric modes. Gommans et al. observed the same polarization dependence for the RBM, G, and D modes from bundled SWCNTs using the polarized Raman spectroscopy [41]. The result is well reproduced with the assumption that only A-symmetry phonon modes are observed. In bundled SWCNTs, it would be difficult to detect non-totally symmetric modes owing to the imperfect alignment of the SWCNT axes. In SWCNTs, the RBM, G, and D modes tend to exhibit the same polarization dependence because the non-totally symmetric modes have a lower Raman intensity than the totally symmetric mode.

6.3 Coherent Phonon

Raman spectroscopy provides energy-domain information, such as the strength of the carrier-phonon interaction. Coherent phonons, on other hand, provide time-domain information, such as the dynamics of the carrier-phonon interaction. Coherent phonons are generated with ultrashort laser pulses and observed with pump-probe techniques [13]. In the pump-probe measurement, the pump pulse creates non-equilibrium carrier and phonon distributions, which modulate the optical response of the sample. A probe pulse, delayed in time by Δt, is used to measure the pump-induced optical response as a function of Δt. The time-domain spectra obtained with pump-probe techniques provide information on the phonon frequency, the phonon dephasing time, and the phase of the phonon oscillations. Furthermore, with time-resolved Fourier transform (FT) analysis, we can track the transient dynamics of carriers and phonons [14, 15, 42]. In the following section, we provide a detailed description of the detection scheme and the experimental apparatus used for the observation of coherent phonons in the SWCNTs in our works.

6.3.1 Detection of Coherent Phonon

Coherent phonons are detected by the time-resolved measurement of optical responses, such as reflectivity (or transmission), four-wave mixing, or second high harmonics. Typically, the time-resolved optical response consists of three components as seen in Fig. 6.6, which shows the time-resolved reflectivity in graphite. First, the initial sharp response originates from an overlap between the pump and probe pulses. Then, the subsequent relaxation of photo-excited carriers produces non-oscillatory exponential decay. Third, the oscillatory signal, which overlaps the electronic response, appears due to the coherent phonons. The oscillatory signal in graphite consists of two components with different vibration periods (21 fs and 1 ps). These components are also identified in the FT spectrum (Fig. 6.6b) and correspond to the inter-layer vibration (E_{2g1}, 42 cm^{-1}) and C-C stretching mode (E_{2g2},

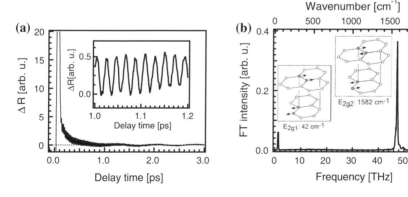

Fig. 6.6 a Time-resolved reflectivity of graphite detected by a sub-10-fs laser pulses, centered at 780 nm. The inset shows the oscillatory signal induced with the G-mode coherent phonons. **b** The FT spectrum of the oscillatory signal in (**a**). The insets show the vibration modes in graphite

$1582\,\mathrm{cm}^{-1}$). The observed frequencies are consistent with the Raman spectrum. The modulation of the reflectivity, originates from the interaction between carriers and phonons. When the atomic displacement (Q) is induced with the phonon oscillation, the electronic state is modified by the electron-phonon interaction. Then, the polarizability at the probe wavelength (χ) is also modulated as a result of the change in the electronic states. Therefore, the reflectivity change (ΔR) is expressed with the first-order Raman tensor $\partial R/\partial Q$,

$$\Delta R = \frac{\partial R}{\partial n} \cdot \Delta n \simeq \frac{\partial R}{\partial \chi} \frac{\partial \chi}{\partial Q} Q \qquad (6.2)$$

where n is the refractive index. The Raman tensor in (6.2) indicates that the selective detection of phonons with a specific symmetry can be performed by changing the laser polarization [43].

Coherent phonons have synchronized motions because ultrashort laser pulses, whose pulse duration is shorter than the vibration period, initiate the oscillation. Then, the in-phase oscillation in coherent phonons produces an observable change in the macroscopic physical parameters, such as the polarizability (i.e., reflectivity), whereas random oscillation (e.g. thermal oscillation), cannot because the out-of-phase oscillation cancels out any change in the optical responses.

6.3.2 Experiment

Here, we briefly mention the experimental set-up we used to detect coherent phonons in SWCNTs [44, 45]. The light source was a Ti-sapphire laser, whose wavelength was centered on 780 nm. The laser pulse duration was sub-10 fs. The laser pulse was split into the pump and probe pulses with a beam splitter. The polarization of both the pump and probe pulses can be controlled identically by using waveplates. The laser intensity was carefully controlled to avoid damaging the sample.

For the selective detection of A or E symmetry phonons in SWCNTs, we used two different detection configurations. To detect all the Raman-active phonon modes in the SWNCTs, we chose a **transient reflectivity** scheme, while for the non-totally symmetric (E_1 and E_2 symmetry) phonon modes, we chose an **anisotropic reflectivity** scheme [44]. In the transient reflectivity measurement, the intensity of the probe beam before and after reflection from the sample was measured with a pair of PIN photodiodes, and the difference between the signal and the reference photocurrents (ΔR) was measured as a function of the pump-probe delay. In the anisotropic reflectivity measurement, the probe beam, which was 45° polarized with respect to the optical plane, was decomposed into polarization components parallel ($R_{//}$) and perpendicular ($R_\perp$) to the optical plane after reflection from the sample [46, 47]. The difference $\Delta R_{\mathrm{aniso}} = R_{//} - R_\perp$ was measured as a function of the pump-probe delay. Since $\Delta R_{\mathrm{aniso}}$ eliminates isotropic responses originating from the isotropic carrier distribution [48] or totally symmetric phonon modes [46, 47], the anisotropic responses originating from the anisotropic carrier distribution or non-totally symmetric phonon modes can be selectively detected. The differential current (ΔR or ΔR_{aniso}) was amplified using a preamplifier with a band-pass filter set to transmit the RBM and the G mode signals simultaneously. The amplified signal was digitized and recorded with a digital oscilloscope. The differential current (ΔR or ΔR_{aniso}) was normalized to $\Delta R/R$ (or $\Delta R_{aniso}/R$) where R is the reflectivity without the pump pulse.

6.4 Anisotropy of Coherent Phonons in Aligned SWCNT

Due to its quasi-one-dimensional structure, SWCNTs exhibit anisotropic optical responses as discussed in Sect. 6.2.1. It is expected that anisotropy can also be found in ultrafast carrier and phonon dynamics because optical responses in SWCNTs, such as absorption [49], photoluminescence [50], photoconductivity [51], and Raman signals [52], depend strongly on the polarization of light with respect to the SWCNT axis. In fact, anisotropic carrier decay is observed in aligned carbon nanotubes [53]. To investigate the polarization dependence in coherent phonon dynamics, we measured the time-resolved reflectivity of the aligned SWCNTs with sub-10-fs laser pulses at 395 nm [44]. In the former part of this section, we present transient and anisotropic reflectivities obtained from aligned SWCNT bundles for the selective detection of coherent phonons with different symmetries. In the latter part, we discuss the different polarization dependence of the RBM and G-mode coherent phonons in terms of the phonon symmetry.

6.4.1 Sample Preparation of Aligned SWCNT

The sample used in this study consisted of SWCNT bundles synthesized by the laser ablation method and purified by reflux in hydrogen peroxide and filtration. A clean hydrophilic glass slide was immersed vertically into an SWCNT/water suspension. Then, a self-assembled SWCNT film was formed on the slide [54]. The direction of the SWCNT axis was evaluated optically, as well as by transmission electron microscopy and polarized Raman spectroscopy [54].

6.4.2 Antenna Effect of the Optical Transition

Due to the antenna effect, the reflectivity from aligned SWCNTs exhibits a polarization dependence [49, 55, 56], and can be used to define the direction of the alignment. The reflectivity of the probe pulse without the pump pulse, is plotted in Fig. 6.7 as a function of the angle (θ) between the laser polarization and the SWCNT axis. In fact, the observed polarization dependence of the reflectivity exhibits a $\cos 2\theta$

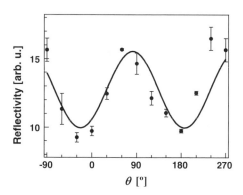

Fig. 6.7 Polarization dependence of the reflectivity without the pump pulse. The *solid circles* are experimental results plotted as a function of the angle (θ) between the laser polarization and the SWCNT axis. The *solid line* is a fitted result with $\cos 2\theta$ (Reprinted with permission from Nano Letters **8**, 3102 (2008). Copyright (2008) American Chemical Society)

dependence. The minimum (maximum), which is observed when the polarization of the laser is parallel (perpendicular) to the SWCNT axis, originates from the antenna (depolarization) effect, as discussed in Sect. 6.2.1.

6.4.3 Transient and Anisotropic Reflectivity Measurements in Aligned SWCNTs

Ultrafast carrier and coherent phonon dynamics, which is excited with an irradiation of the pump pulse, is monitored via the delay-dependent change in the reflectivity of, the probe pulse as shown in Fig. 6.8. Figure 6.8a, b show the time-resolved reflectivity of SWCNTs obtained by the transient and anisotropic reflectivity measurements, respectively. In both measurements, the photo-induced reflectivity consists of three components as seen in graphite. At $\Delta t = 0$, an overlap between the pump and probe pulses exhibits the sharp and intense response, which is followed with the subsequent relaxation of excited carriers. Then, the oscillatory signal appears as a result of the coherent lattice vibrations, that is, coherent phonons.

To extract the oscillation frequencies, the FT spectra of the oscillatory $\Delta R/R$ (Fig. 6.8c) and $\Delta R_{aniso}/R$ (Fig. 6.8d) are obtained after the subtraction of the non-oscillatory carrier responses from ΔR and ΔR_{aniso}. The FT spectrum obtained with the transient reflectivity measurement (Fig. 6.8c) shows that the coherent phonons in SWCNTs can be separated into three frequency regions (1) low-frequency ($>10\,$THz), (2) $1350\,$cm^{-1} ($40\,$THz), and (3) 1560–$1580\,$cm^{-1} ($47\,$THz), while that obtained by the anisotropic reflectivity measurement (Fig. 6.8d) shows only the highest frequency component.

The multiple peaks in the low-frequency region correspond to the RBMs of SWC-NTs with different diameters (Fig. 6.8c). Because they are totally symmetric [19, 20, 26], RBMs in aligned SWCNTs can be observed only in the transient reflectivity. The dependence of the RBM frequency on the SWCNT radius [28] reveals that the observed signal originates from SWCNTs with diameters of 1.0(1), 1.4(2), and

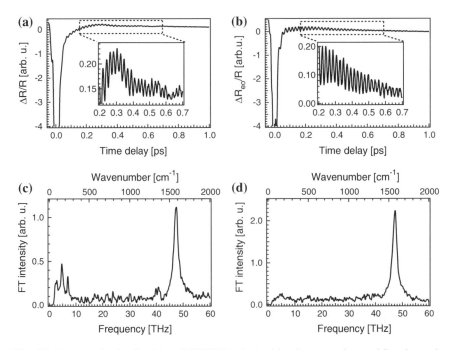

Fig. 6.8 Time-resolved reflectivity of SWCNTs obtained by the **a** transient and **b** anisotropic reflectivity measurements. The inset is the enlarged signal for the oscillatory components. The FT spectrum of the time-resolved reflectivity of SWCNTs obtained by the **c** transient, and **d** anisotropic reflectivity measurements (Reprinted with permission from Nano Letters **8**, 3102 (2008). Copyright (2008) American Chemical Society)

1.6(2) nm. According to the Kataura plot [22], the SWCNTs are resonantly excited through the E_{22}^S, E_{11}^M, E_{33}^S, and E_{44}^S transitions with the irradiation of 395 nm laser pulses.

The peak at 40 THz can be observed only with the transient reflectivity (Fig. 6.8c). It is assigned as the D mode on the basis of the Raman spectra of the SWCNTs. The D-mode coherent phonons are observed only with the transient reflectivity measurements probably due to the isotropic distribution of defects in the SWCNTs, which produces an isotropic optical response. The observation of D-mode coherent phonons indicates that even a phonon with a non-zero wavevector ($q \neq 0$) takes coherent lattice oscillations [57]. The present result suggests that the synchronized motion of phonons whose phases are locked could generate coherent phonons.

The peak at 47 THz is assigned to the G mode on the basis of the Raman spectra of the SWCNTs. Since the G mode consists of totally and non-totally symmetric phonon modes [20, 26], it can be observed with both transient and anisotropic reflectivity measurements. As discussed in Sect. 6.2.2, the G-mode phonons consist of longitudinal and transverse modes with different symmetries and so exhibit multiple FT peaks. However, the short dephasing time of the G-mode coherent phonons in this

Fig. 6.9 FT spectrum of the time-resolved reflectivity of SWCNTs when both the pump and probe polarizations are parallel (*black*) or perpendicular (*gray*) to the SWCNT axis (Reprinted with permission from Nano Letters **8**, 3102 (2008). Copyright (2008) American Chemical Society)

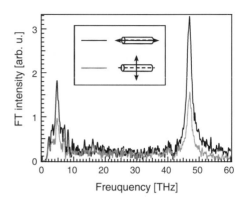

sample (less than 1 ps) make it difficult to resolve G-mode coherent phonons with different frequencies.

6.4.4 Antenna Effect of Coherent Phonons

Because of the antenna effect, the phonon amplitudes of all modes depend on the laser polarization with respect to the SWCNT axis. When both the pump and the probe polarizations are parallel to the SWCNT axis, both the RBM and G modes have higher intensities than when they are perpendicular as shown in Fig. 6.9. As observed in resonant Raman spectra [41], Fig. 6.9 indicates that the antenna effect also dominates the amplitude of coherent phonons.

6.4.5 Polarization Dependence of Coherent Phonons

To investigate the polarization dependence of coherent phonons in SWCNTs, we measured the transient reflectivity as a function of the angle between the polarization of the pump laser and the SWCNT axis (θ_1). Figure 6.10 shows the transient reflectivity as a function of θ_1 whereas the polarization of the probe pulse was perpendicular to the SWCNT axis. In Fig. 6.10, the amplitude of the G-mode phonons is nearly zero at $\theta_1 = -30°$ and $30°$ while the RBMs appear at all angles. The FT spectra are shown in Fig. 6.11a to clarify the polarization dependence of each oscillatory component. The amplitude of the G mode exhibits maxima at $\theta_1 = -90°$ and $90°$ minima at $\theta_1 = -30°$ and $30°$. Because of the one-dimensional structure of SWC-NTs, the polarization dependence has mirror symmetry at $\theta_1 = 0°$ [52, 58]. However, the degradation of the sample during the measurement leads to the imperfect mirror symmetry at $\theta_1 = 0°$ in the FT spectra. To remove this artifact, the ratio between the integrated intensities of the G mode and those of the RBM, I_G/I_{RBM}, is plotted

Fig. 6.10 Time-resolved transient reflectivity obtained by rotating the polarization of the pump pulse (*red*). The polarization of the probe pulse (*blue*) is set perpendicular to the SWCNT axis (Reprinted with permission from Nano Letters **8**, 3102 (2008). Copyright (2008) American Chemical Society)

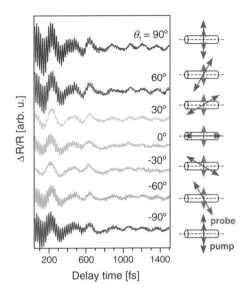

as a function of $\theta_1 - \theta_2$ (filled circles in Fig. 6.11b), where θ_2 is the polarization angle between the probe pulse and the SWCNT axis. The intensity, I_G/I_{RBM}, has its maximum at $\theta_1 - \theta_2 = 0°$ and $180°$, falls almost to zero at $\theta_1 - \theta_2 = 60°$ and $120°$, and has a local maximum at $\theta_1 - \theta_2 = 90°$. For comparison, the

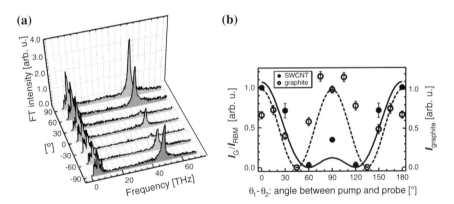

Fig. 6.11 a FT spectrum of oscillatory signals in Fig. 6.10. The color of each plot corresponds to the FT spectrum with the same color as that in Fig. 6.10. **b** *Solid circles* are the ration of the integrated intensity of the G mode and RBM, I_G/I_{RBM}, as a function of the angle between the pump and probe pulses. The *solid line* is the fitted result with $|\alpha + \beta \cos 2(\theta_1 - \theta_2)|^2$ where $\alpha/\beta = 0.5$. For comparison, the open circles show the polarization dependence of graphite as a function of the angle between the pump and probe pulses. The *dotted line* is the fitted results for graphite with $|\cos 2(\theta_1 - \theta_2)|^2$ (Reprinted with permission from Nano Letters **8**, 3102 (2008). Copyright (2008) American Chemical Society)

polarization dependence of graphite obtained with the transient reflectivity measurement is plotted with open circles in Fig. 6.11b. As expected from the E_{2g} Raman tensor of the G mode in graphite, the intensity shows a $|\cos 2(\theta_1 - \theta_2)|^2$ dependence [42, 43]. The different polarization dependences of graphite and SWCNTs indicate that phonon modes, which become Raman active due to zone folding, should result in the polarization dependence of the SWCNTs.

To characterize the polarization dependence of the SWCNTs further, we performed the measurements on angles θ_1 and θ_2. Figure 6.12 shows I_G/I_{RBM} as a function of (θ_1, θ_2). In Fig. 6.12, I_G/I_{RBM} depends strongly on the relative angle, $\theta_1 - \theta_2$. When the pump and probe polarizations are parallel, that is $|\theta_1 - \theta_2| = 0°$ or 180°, I_G/I_{RBM} is maximum, and when the relative angle between the pump and probe $|\theta_1 - \theta_2|$ is 60° or 120°, I_G/I_{RBM} is zero (indicated by dotted lines in Fig. 6.12). Furthermore, I_G/I_{RBM} has a local maximum when the pump and probe are perpendicular, that is, $|\theta_1 - \theta_2| = 90°$.

To discuss the observed G mode polarization dependence in Figs. 6.11 and 6.12, we take account of the superposition of G-mode phonons with different symmetries. Such a superposition has been reported for GaAs/AlGaAs multiple quantum wells [59], in which the summation of the corresponding Raman tensors produces the characteristic polarization dependence.

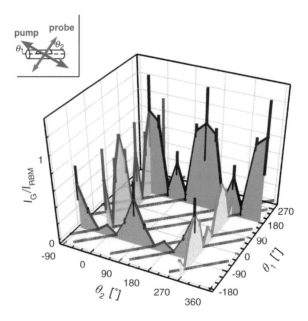

Fig. 6.12 Polarization dependence of I_G/I_{RBM} plotted as a function of (θ_1, θ_2) where $\theta_1(\theta_2)$ is the angle of the pump (probe) polarization with respect to the SWCNT axis. *Dotted lines* connect between points where I_G/I_{RBM} nearly equals zero (Reprinted with permission from Nano Letters **8**, 3102 (2008). Copyright (2008) American Chemical Society)

In CNTs, when the excitation light propagates in the x direction with a polarization vector in the z direction corresponding to the nanotube axis, C_6 symmetry Raman tensors with A, E_1, and E_2 symmetries in the yz plane can be expressed as

$$A : \begin{pmatrix} a & 0 \\ 0 & b \end{pmatrix}; \quad E_1 : \begin{pmatrix} 0 & c \\ c & 0 \end{pmatrix}; \quad E_2 : \begin{pmatrix} d & 0 \\ 0 & 0 \end{pmatrix} \tag{6.3}$$

respectively (Table 1 in [60]). The scattering efficiencies of the A, E_1, and E_2 symmetries are given by

$$A : |\frac{a+b}{2} + \frac{b-a}{2} \cos(2\theta)|^2, \tag{6.4}$$

$$E_1 : |c \cos(2\theta)|^2, \tag{6.5}$$

$$E_2 : |d \frac{1 - \cos(2\theta)}{2}|^2, \tag{6.6}$$

respectively, where θ is the pump polarization [43, 59]. The polarization dependence of the E_{2g} symmetry phonon in graphite is well reproduced with $|\cos(2\theta)|^2$. A previous polarized Raman measurement in CNTs shows experimentally that the G mode with A symmetry has $a = 0$ [41]. Then, the scattering efficiency of the A symmetry phonon is $|\frac{b}{2}(1 + \cos(2\theta))|^2$. Neither the A nor the E symmetry Raman tensors alone can explain the present polarization dependence of CNTs with minima at $\theta = 60°$ and $120°$.

When all of the contributions from the A, E_1, and E_2 symmetries are taken into consideration, the total scattering intensity is given by

$$\frac{dS}{d\Omega} \propto |\alpha + \beta \cos(2\theta)| \tag{6.7}$$

where α and β combine the elements of the Raman tensors. Indeed, the observed polarization dependence in CNTs can be well reproduced by (6.7) provided that $\alpha/\beta = 0.5$ (solid line in Fig. 6.11b). The G-mode polarization dependence can be explained by the superposition of A, E_1, and E_2 symmetry Raman tensors.

In a previous polarized Raman measurement with linearly and circularly polarized light [61], the phonon symmetries of CNTs were investigated by measuring Raman tensor invariants. In this work, the ratio between the asymmetric and the isotropic invariant shows that the Raman peaks of CNTs are not due to modes with distinctly different symmetries but are due to a superposition of the A and E symmetry modes. Because our result also indicates the superposition of the A and E symmetry modes, it corresponds to a real-time observation of the superposition of phonons with different symmetries. The disappearance of the total G-mode phonon amplitude at a certain angle in our time domain data suggests that G-mode phonons with different symmetries might oscillate with different phases.

6.5 Coherent Phonons in Metallic SWCNTs

To obtain information from SWCNTs with a specific chirality, we generally use the resonance effect. Through the resonance effect, we can selectively excite the SWCNT, whose transition energy matches to the excitation photon energy [19, 20]. However, we cannot use this scheme for time-resolved measurements. In time-resolved measurements, we need to use ultrashort laser pulses, which have a wide energy dispersion due to the uncertainty principle. The broad energy distribution, which allows the simultaneous excitation of SWCNTs with different symmetries, makes it difficult to obtain the detailed information about SWCNTs with the specific chirality because most SWCNT sample contain a broad distribution of tubes with different chiralities [16, 44, 62]. Therefore, we need less dispersed samples, which could be obtained by a particular preparation scheme [63], or a separation scheme [64, 65]. In this section, we report the time-resolved reflectivity in metallic SWCNTs, that were extracted from mixed samples [45]. In previous studies, coherent phonons have been observed for semiconducting SWCNTs [16, 17, 66], but not for metallic SWCNTs although measurements were performed in mixed samples [14]. Here, we describe our observation of ultrafast carrier and coherent phonons in metallic SWCNTs.

6.5.1 Experiment

Samples enriched with metallic SWCNTs were obtained with density gradient ultracentrifugation [65] from bundled SWCNTs produced with an arc discharge (Meijo Nano Carbon Co.). An original sample before separation exhibited two absorption peaks at around 700 and 1000 nm, which correspond to the first excited state of metallic SWCNTs and the second excited state of semiconducting SWCNTs, respectively (the inset in Fig. 6.13). The sample after the separation exhibited a single peak at around 700 nm. The ratio between metallic and semiconducting SWCNTs was estimated to be 3:1 from the absorption intensities. To observe coherent phonons in metallic SWCNTs, we performed a time-resolved reflectivity measurement with sub-10-fs laser pulses, whose wavelengths ranged from 690 to 850 nm. The intensity of the pump pulse varied from 1.1×10^{10} to 2.2×10^{10} W/cm^2. We measured the anisotropic reflectivity, which reduced the isotropic carrier response from the SWCNTs [42] and gave a better signal to noise ratio for the observation of coherent phonons.

6.5.2 Results

The anisotropic reflectivity change of the metallic SWCNTs is shown in Fig. 6.13. Following an initial sharp and intense peak at $\Delta t = 0$ caused by an overlap between

Fig. 6.13 Anisotropic time-resolved reflectivity of metallic SWCNTs. The pump laser intensity was 2.2×10^{10} W/cm^2. The inset shows the absorption spectra for the mixed sample (*dotted*) and metallic SWCNTs (*solid*)

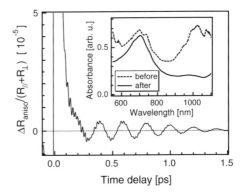

the pump and probe pulses [14, 42, 44], the non-oscillatory signal appears and decays within 30 fs, which is much shorter then the exciton decay time (sub ps) [67]. According to the previous time-resolved photoelectron spectroscopy, the observed decay is due to the scattering among free carriers [68]. After the carrier response, the signal exhibits damped oscillatory behavior induced by coherent lattice vibrations. We extracted the phonon frequency from the oscillatory signal using FT analysis. Figure 6.14 shows the FT power spectra. These spectra have three frequency regions that can be assigned to the RBM (5 THz), D mode (39 THz), and G mode (47 THz). The random distribution of the SWCNT axes in the sample allows the detection of these three phonon modes with the anisotropic reflectivity measurement.

As regards the RBM, we obtained two frequencies (4.89 and 4.75 THz) at the minimum pump laser intensity. The relationship between the RBM frequency and the diameter [28] indicates that the observed signal originates from SWCNTs with diameters of 1.49 and 1.54 nm. According to the Kataura plot [22], both SWCNTs, which

Fig. 6.14 Power spectra of the oscillatory signal in Fig. 6.14. The inset shows an enlargement of the spectrum around 47 THz. To resolve the split structure of the G modes, we set the band pass filter to cut the signal with the RBM signal. The *dotted lines* are fitted components for the G-mode spectra; a BWF lineshape and three Lorentzians

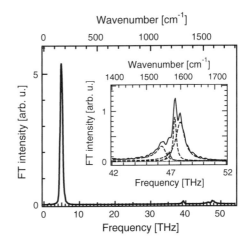

are resonantly excited under the present experimental conditions, are determined as metallic. Note that coherent phonons are observed in metallic SWCNTs.

The G mode, which clearly exhibits multiple peaks around 47 THz due to the long dephasing time, is decomposed into four components (shown by dotted lines in the inset in Fig. 6.14). The component with the lowest frequency has the BWF line shape and the other three are Lorentzians. The BWF line shape, which is due to the coupling of a discrete phonon with an electronic continuum [35, 36], also confirms the observation of coherent phonons in metallic SWCNTs. As mentioned in Sect. 6.2.2, the peaks in the G mode are analyzed as two modes, i.e., one labeled as the G^- mode with a lower frequency (ω_{G^-}) and the other as the G^+ mode (ω_{G^+}) with a higher frequency. According to the previous Raman measurements [33], the difference between ω_{G^-} and ω_{G^+} for metallic SWCNTs depends on the diameters (d_t) as follows,

$$\omega_{G^+} - \omega_{G^-} \text{ [THz]} = \frac{2.39}{d_t^2} \text{ [nm]} \qquad (6.8)$$

With (6.8), $\omega_{G^+} - \omega_{G^-}$ is calculated at 1.01–1.08 THz for SWCNTs with diameters of 1.54–1.49 nm. In fact, the observed difference between the lower (46.5 and 47.0 THz) and higher frequencies (47.5 and 48.0 THz) ranges from 1.0 to 1.5 THz, which is comparable to the calculated values. Therefore, we can conclude that we observed the G^- and G^+-mode coherent phonons, corresponding to the longitudinal and transverse optical phonons in metallic SWCNTs, respectively. The less disperse sample provides us clear information about SWCNTs with the specific chirality and also allows us the detailed study for each phonon mode.

6.6 Summary

We have presented time-resolved studies for ultrafast coherent phonon dynamics in SWCNTs. The time-resolved observation of coherent phonons is basically consistent with the spectrum obtained with conventional Raman spectroscopy, but also provides additional information.

In the aligned SWCNTs, we have selectively observed totally symmetric modes (the RBM, D, and G modes) or asymmetric modes (only the G mode) by choosing the transient or anisotropic reflectivity measurements. We have observed the polarization dependence of both the reflectivity and the amplitude of coherent phonons induced by the antenna effect. The polarization dependence of the G-mode coherent phonons is explained with the superposition of phonon modes with different symmetries. Coherent phonon has the possibility for the selective detection of phonons with different symmetries in SWCNTs.

By using metallic SWCNTs, extracted from a mixed sample, we succeeded in observing ultrafast carrier and phonon dynamics. The free carriers in metallic SWC-NTs decay within 30 fs through the carrier-carrier scattering. We have identified the RBM, D-mode and G-mode coherent phonons. We also resolved G^-- and G^+-mode coherent phonons, which correspond to the longitudinal and transverse optical phonons. The BWF lineshape in the G^- mode, originating from the coupling between coherent phonons and carriers, also supports the observation of coherent phonons in metallic SWCNTs. With the less disperse sample, we succeeded in observing the detailed carrier and phonon dynamics in metallic SWCNTs.

Coherent phonon is a powerful tool for accessing the fundamental properties of a material, but also has the potential to control the material. Dumitrica et al. theoretically pointed out that we can open nanotube caps by exciting coherent phonons with a specific mode in SWCNTs [69, 70]. The phase of terahertz wave emitted from SWCNTs can be also controlled with a pair of laser pulses [71]. In combination with the laser-pulse shaping technique, we believe the coherent phonon will open new areas of physics and technology in solids.

6.7 Acknowledgment

K. K. thanks Prof. Saito and Dr. Sasaki for discussions about the results, Prof. Petek, Dr. Ishioka, Dr. Jie for the experiments on aligned SWCNTs, Dr. Kataura, Dr. Miyata, and Dr. Yanagi for the sample preparation of metallic SWCNTs, Dr. Nakano, Dr. Sogawa, Dr. Ishizawa, and Dr. Gotoh for the experiments on metallic SWCNTs. K. K. would also like to thank Prof. Shudo for his help with the figures, Prof. Hase, Prof. Misochko, and Dr. Kubo for general discussions about coherent phonons. This research was supported by Grant-in-Aid for Scientific Research (KAKENHI) (No. 20750021, 23104515, 21310065) from the Japan Society for the Promotion of Science. Drawings in Figs. 6.1, 6.3, 6.5, 6.8, 6.10, and 6.12 are reprinted from [72] under permission of Vacuum Society of Japan.

References

1. Z. Yao, C.L. Kane, C. Dekker, Phys. Rev. Lett. **84**, 2941–2944 (2000)
2. A. Javey, J. Guo, M. Paulsson, Q. Wang, D. Mann, M. Lundstrom, H. Dai, Phys. Rev. Lett. **92**, 106804 (2004)
3. K.S. Novoselov, A.K. Geim, S.V. Morozov, D. Jiang, Y. Zhang, S.V. Dubonos, I.V. Grigorieva, A.A. Firsov, Science **306**, 666–669 (2004)
4. X. Du, I. Skachko, A. Barker, E.Y. Andrei, Nat. Nanotech. **3**, 491–495 (2008)
5. R. Saito, M. Fujita, G. Dresselhaus, M.S. Dresselhaus, App. Phys. Lett. **60**, 2204–2206 (1992)
6. S.J. Tans, A.R.M. Verschueren, C. Dekker, Nature **393**, 49–52 (1998)
7. T. Hertel, G. Moos, Phys. Rev. Lett. **84**, 5002–5005 (2000)
8. S. Tatuura, M. Furuki, Y. Sato, I. Iwasa, M. Tian, H. Mitsui, Adv. Mater. **15**, 534–537 (2003)
9. N. Kamaraju, S. Kumar, Y.A. Kim, T. Hayashi, H. Muramatsu, M. Endo, A.K. Sood, Appl. Phys. Lett. **95**, 081106 (2009)
10. F. Wang, A.G. Rozhin, V. Scardaci, Z. Sun, F. Hennrich, I.H. White, W.I. Milne, A.C. Ferrari, Nat. Nanotech. **3**, 738–742 (2008)
11. Y.-W. Song, S. Yamashita, S. Maruyama, Appl. Phys. Lett. **92**, 021115 (2008)

12. T.R. Schibli, K. Minoshima, H. Kataura, E. Itoga, N. Minami, S. Kazaoui, K. Miyashita, M. Tokumoto, Y. Sakakibara, Opt. Express **13**, 8025–8031 (2005)
13. T. Dekorsy, G.C. Cho, H. Kurz, *Light Scattering in Solids VIII Fullerenes, Semiconductor Surfaces, Coherent Phonons* (Springer, Berlin, 2000)
14. A. Gambetta, C. Manzoni, E. Menna, M. Meneghetti, G. Cerullo, G. Lanzani, S. Tretiak, A. Piryatinski, A. Saxena, R.L. Martin, A.R. Bishop, Nat. Phys. **2**, 515–520 (2006)
15. M. Hase, M. Kitajima, A.M. Constantinescu, H. Petek, Nature **426**, 51–54 (2003)
16. J.-H. Kim, K.-J. Han, N.-J. Kim, K.-J. Yee, Y.-S. Lim, G.D. Sanders, C.J. Stanton, L.G. Booshehri, E.H. Hároz, J. Kono, Phys. Rev. Lett. **102**, 037402 (2009)
17. L. Läer, C. Badermaier, J. Crochet, T. Hertel, D. Brida, G. Lanzani, Phys. Rev. Lett. **102**, 127401 (2009)
18. G.D. Sanders, C.J. Stanton, J.-H. Kim, Y.-S. LimE, H. Hároz, L.G. Booshehri, J. Kono, R. Saito, Phys. Rev. B **79**, 205434 (2009)
19. S. Reich, C. Thomsen, J. Maultzsch, *Carbon Nanotubes* (Wiley-VCH Verlag GmbH & Co. KGaA, Weinheim, 2004)
20. R. Saito, G. Dresselhaus, M.S. Dresselhaus, *Physical Properties of Carbon Nanotubes* (Imperial College Press, London, 1998)
21. R. Sato, OYO BUTSURI 70, pp. 1196–1199 (2001) [in Japanese]
22. H. Kataura, Y. Kumazawa, Y. Maniwa, I. Umezu, S. Suzuki, Y. Ohtsuka, Y. Achiba, Synth. Met. **103**, 2555–2558 (1999)
23. A.M. Rao, E. Richter, S. Bandow, B. Chase, P.E. Eklund, K.A. Williams, S. Fang, K.R. Subbaswamy, M. Menon, A. Thess, R.E. Smalley, G. Dresselhaus, M.S. Dresselhaus, Science **275**, 187–191 (1997)
24. Y. Wang, K. Kempa, K. Kimball, J.B. Carlson, G. Benham, W.Z. Li, T. Kempa, J. Rybczynski, A. Herczynski, Z.F. Ren, Appl. Phys. Lett. **85**, 2607–2609 (2004)
25. H. Ajiki, T. Ando, Phys. B **201**, 349–352 (1994)
26. R. Saito, T. Takeya, T. Kimura, G. Dresselhaus, M.S. Dresselhaus, Phys. Rev. B **57**, 4145–4153 (1998)
27. M.S. Dresselhaus, P.C. Eklund, Adv. Phys. **49**, 705–814 (2000)
28. S.M. Bachilo, M.S. Strano, C. Kitterell, R.H. Hauge, R.E. Smalley, R.B. Weisman, Science **298**, 2361–2366 (2002)
29. A. Jorio, G. Dresselhaus, M.S. Dresselhaus, M. Souza, M.S.S. Dantas, M.A. Pimenta, A.M. Rao, R. Saito, C. Liu, H.M. Chang, Phys. Rev. Lett. **85**, 2617–2619 (2000)
30. A. Jorio, M.A. Pimenta, A.G.S. Filho, GeG Samsonidze, A.K. Swan, M.S. Ünlü, B.B. Goldbergh, R. Saito, G. Dresselhaus, M.S. Dresselhaus, Phys. Rev. Lett. **90**, 107403 (2003)
31. R. Saito, A. Jorio, J.H. Hafner, C.M. Lieber, M. Hunger, T. McClure, G. Dresselhaus, M.S. Dresselhaus, Phys. Rev. B **64**, 085312 (2001)
32. A. Kasuya, Y. Sasaki, Y. Saito, K. Tohji, Y. Nishina, Phys. Rev. Lett. **78**, 4434–4437 (1997)
33. A. Jorio, A.G. Souza Filho, G. Dresselhaus, M.S. Dresselhaus, A.K. Swan, M.S. Ünlü, B.B. Goldberg, M.A. Pimenta, J.H. Hafner, C.M. Lieber, R. Saito, Phys. Rev. B **65**, 155412 (2002)
34. S. Piscanec, M. Lazzeri, J. Robertson, A.C. Ferrari, F. Mauri, Phys. Rev. B **75**, 035427 (2007)
35. S.D.M. Brown, A. Jorio, P. Corio, M.S. Dresselhaus, G. Dresselhaus, R. Saito, K. Kneipp, Phys. Rev. B **63**, 155414 (2001)
36. A.M. Rao, P.C. Eklund, S. Bandow, A. Thess, R.E. Smalley, Nature **388**, 257–259 (1997)
37. F. Tuinstra, J.L. Koenig, J. Phys. Chem. **53**, 1126–1130 (1970)
38. C. Thomsen, S. Reich, Phys. Rev. Lett. **85**, 5214–5217 (2000)
39. R. Saito, A. Jorio, A.G.S. Filho, G. Dresselhaus, M.S. Dresselhaus, M.A. Pimenta, Phys. Rev. Lett. **88**, 027401 (2002)
40. R. Saito, TANSO **205**, 276–284 (2002)
41. H.H. Gommans, J.W. Alldredge, H. Tashiro, J. Park, J. Magnuson, A.G. Rinzler, J. Appl. Phys. **88**, 2509–2514 (2000)
42. K. Ishioka, M. Hase, M. Kitajima, L. Wirtz, A. Rubio, H. Petek, Phys. Rev. B **77**, 121402(R) (2008)
43. K.J. Yee, K.G. Lee, E. Oh, D.S. Kim, Y.S. Lim, Phys. Rev. Lett. **88**, 105501 (2002)

44. K. Kato, K. Ishioka, M. Kitajima, J. Tang, R. Saito, H. Petek, Nano Lett. **8**, 3102–3108 (2008)
45. K. Kato, K. Oguri, A. Ishizawa, H. Gotoh, H. Nakano, T. Sogawa, Appl. Phys. Lett. **97**, 121910 (2010)
46. T. Dekorsy, H. Auer, C. Waschke, H.J. Bakker, H.G. Roskos, H. Kurz, V. Wagner, P. Grosse, Phys. Rev. Lett. **74**, 738–741 (1995)
47. M. Hase, K. Mizoguchi, H. Harima, S. Nakashima, M. Tani, K. Sakai, M. Hangyo, Appl. Phys. Lett. **69**, 2474–2476 (1996)
48. A.J. Sabbah, D.M. Riffe, Phys. Rev. B **66**, 165217 (2002)
49. Y. Murakami, E. Einarsson, T. Edamura, S. Maruyama, Phys. Rev. Lett. **94**, 087402 (2005)
50. J. Lefebrvre, P. Finnie, Phys. Rev. Lett. **98**, 167406 (2007)
51. M. Freitag, Y. Martin, J.A. Misewich, R. Martel, Ph Avouris, Nano Lett. **3**, 1067–1071 (2003)
52. A. Jorio, A. G. S. Filho, V. W. Brar, A. K. SwanM. S. Ünlü, B. B. Goldber, A. Righi, J. H. Hafner, C. M. Lieber, R. Saito, G. Dresselhaus, M. S. Dresselhaus, Phys. Rev. B **65**, 121402(R) (2002)
53. Y. Hashimoto, Y. Murakami, S. Maruyama, J. Kono, Phys. Rev. B **75**, 245408 (2007)
54. H. Shimoda, S.J. Oh, H.Z. Geng, R.J. Walker, X.B. Zhang, L.E. MeNeil, O. Zhou, Adv. Mater. **14**, 899–901 (2002)
55. J. Hwang, H.H. Gommans, A. Ugawa, A. Tashiro, R. Haggenmueller, K.I. Winey, J.E. Fischer, D.B. Tanner, A.G. Rinzler, Phys. Rev. B **62**, R13310–R13313 (2000)
56. M.F. Islam, D.E. Milkie, C.L. Kane, A.G. Yodh, J.M. Kikkawa, Phys. Rev. Lett. **93**, 037404 (2004)
57. A.V. Kuznetsov, C.J. Stanton, Phys. Rev. Lett. **73**, 3243–3246 (1994)
58. J. Kim, J. Park, B.Y. Lee, D. Lee, K. Yee, Y. Lim, L.G. Booshehri, E.H. Haroz, J. Kono, S. Baik, J. Appl. Phys. **105**, 103506 (2009)
59. K.J. Yee, Y.S. Lim, T. Dekorsy, D.S. Kim, Phys. Rev. Lett. **86**, 1630–1633 (2001)
60. R. Loudon, Adv. Phys. **50**, 813–864 (2001)
61. S. Reich, C. Thomsen, G. S. Duesberg, S. Roth, Phys. Rev. B **63**, 041401(R) (2001)
62. Y.-S. Lim, K.-J. Yee, J.-H. Kim, E.H. Haroz, J. Shaber, J. Kono, S.K. Doorn, R.H. Hauge, R.E. Smalley, Nano Lett. **6**, 2696–2700 (2006)
63. S.M. Bachilo, L. Balzano, J.E. Herrera, F. Pompeo, D.E. Resasco, R.B. Weisman, J. Am. Chem. Soc. **2003**, 11186–11187 (2003)
64. M.S. Arnold, A.A. Green, J.F. Hulvat, S.I. Stupp, M.C. Hersam, Nat. Nanotech. **1**, 60–65 (2006)
65. K. Yanagi, Y. Miyata, H. Kataura, Appl. Phys. Exp. **1**, 034003 (2008)
66. K. Makino, A. Hirano, K. Shiraki, Y. Maeda, M. Hase, Phys. Rev. B **80**, 245428 (2009)
67. T. Hertel, R. Fasel, G. Moos, Appl. Phys. A **75**, 449–465 (2002)
68. O.J. Korovyanko, C.-X. Sheng, Z.V. Vardeny, A.B. Dalton, R.H. Baughman, Phys. Rev. Lett. **92**, 017403 (2004)
69. T. Dumitrică, M.E. Garcia, H.O. Jeschke, B. Yakobson, Phys. Rev. Lett. **92**, 117401 (2004)
70. T. Dumitrică, M.E. Garcia, H.O. Jeschke, B.I. Yakobson, Phys. Rev. B **74**, 193406 (2006)
71. R.W. Newson, J.-M. Ménard, C. Sames, M. Betz, H.M. van Driel, Nano Lett. **8**, 1586–1589 (2008)
72. K. Kato, M. Kitajima, J. Vac. Soc. Jpn. **53**, 317–326 (2010)

Chapter 7
Generation and Observation of GHz–THz Acoustic Waves in Thin Films and Microstructures Using Optical Methods

Osamu Matsuda and Oliver B. Wright

7.1 Introduction

Acoustic waves are scattered by structural inhomogeneities. This scattering has been used to investigate the internal structure of media, allowing the nondestructive testing of airplane bodies, the development of underwater sonar, and the ultrasonic inspection of internal organs, for example. All these methods, using acoustic waves in the frequency range from $\sim 10\,$Hz to $\sim 10\,$MHz, have a fundamental limitation for the spatial resolution $\sim 0.1\,$mm to $\sim 100\,$m set by wave diffraction.

To image microscopic structures, one therefore needs to use acoustic waves with much higher frequencies. Commonly used ultrasonic transducers based on piezoelectric devices are, however, not efficient for generating acoustic waves with a frequency higher than 1 GHz. Acoustic waves in the GHz to THz range can be accessed with ultrafast optical techniques. Advances in laser technology have made it possible to produce light pulses with sub-ps temporal duration with ease. Bulk acoustic waves with frequency components up to 1 THz, for example, can be generated by a quasi-instantaneous and localized stress distribution in a medium caused by the absorption of such ultrashort light pulses. The propagation of the generated acoustic pulses can be monitored with THz bandwidth also using ultrashort light pulses. Terahertz acoustic waves in solids have wavelengths in the nanometer range, and these can be used to inspect nanoscale structure inside solid media.

The acoustic wave velocity is much slower (approximately by a factor of 10^{-5}) than that of light. Therefore, at the same frequency the wavelength of acoustic waves is $\sim 10^{-5}$ that of electromagnetic waves. This has had ramifications in telecom-

O. Matsuda (✉) · O. B. Wright
Division of Applied Physics, Faculty of Engineering, Hokkaido University,
Sapporo060-8628, Japan
e-mail: omatsuda@eng.hokudai.ac.jp

O. B. Wright
e-mail: assp@eng.hokudai.ac.jp

K. Shudo et al. (eds.), *Frontiers in Optical Methods*,
Springer Series in Optical Sciences 180, DOI: 10.1007/978-3-642-40594-5_7,
© Springer-Verlag Berlin Heidelberg 2014

munication technology by allowing the size of filters, resonators, and waveguides for GHz signal processing to be shrunk. These devices exploit acoustic waves with a wavelength of a few microns propagating along solid surfaces (surface acoustic waves, SAWs). Ultrafast optical methods are also useful for the investigation of the propagation of GHz-SAWs, since the acoustic field can be imaged in the time domain with micron spatial resolution in a noncontact and nondestructive manner.

In this article, we review recent work on the generation and detection of very high-frequency acoustic waves using ultrashort light pulses, mainly concentrating on our own results. In Sect. 7.2, we present the principles of the measurement techniques. Results concerning sub-THz bulk acoustic waves are presented in Sect. 7.3, and results concerning GHz-SAW imaging are presented in Sect. 7.4.

7.2 Picosecond Laser Ultrasonics

In the 1980s, Thomsen et al. [1, 2] found that the irradiation of ps laser pulses on thin opaque amorphous semiconductor films of a few hundred nanometers in thickness caused a transient and periodic optical reflectivity variation with a period of ∼10 ps. Each absorbed light pulse generated a longitudinal acoustic pulse with a spatial width of several tens of nanometers near the sample surface, and this acoustic pulse propagated back and forth inside the film. The optical reflectivity variation is caused by the arrival of acoustic pulses at the film surface. From the period of the reflectivity variation it is possible to measure the film thickness or the longitudinal sound velocity. Since the work of Thomsen et al. similar techniques have been applied to metals [2–28], semiconductors [1, 2, 29–38] and dielectrics [1, 3, 39–48] in the form of mono- or multi-layer films. These techniques have been also applied to liquids [49–52], magnetic materials [53–57], quasi-crystals [58], quantum wells or superlattices [59–73], nano- or micro-structures [74–80], interfaces [21, 81], and biological cells [82, 83]. Acoustic solitons have also been excited and detected [84–88]. The general research field, now well developed, is known as picosecond laser ultrasonics (or simply picosecond ultrasonics). The experimental methods for picosecond laser ultrasonics will be briefly reviewed in the following sub-sections.

7.2.1 Basic Experimental Setup

Figure 7.1 shows a typical experimental setup for a picosecond laser ultrasonics measurement based on reflectivity variations. This makes use of the optical pump-probe technique. A Ti-sapphire laser generates light pulses of wavelength 780–850 nm, duration ∼100 fs, repetition rate ∼80 MHz, and energy ∼10 nJ/pulse. A beam of light pulses, known as the pump beam, is focused with a diameter of ∼1–10 μm onto a sample surface. The pump light, typically having an energy per pulse ∼0.1–1 nJ, can generate an acoustic pulse, with a strain amplitude typically in the

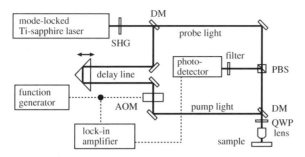

Fig. 7.1 Schematic diagram of a typical setup for picosecond laser ultrasonics measurement. *SHG* second harmonic generation crystal, *AOM* acousto-optic modulator, *QWP* quarter wave plate, *pol.* polarizer, *PBS* polarizing beam splitter, *DM* dichroic mirror

range 10^{-6} to 10^{-4}, by a variety of mechanisms, such as by thermal expansion, by the deformation potential mechanism or by piezoelectric effects for example. The propagating acoustic waves modulate the optical reflectivity in the picosecond time regime. A second light beam from the laser is converted to the second harmonic using a nonlinear optical crystal (β-BaB$_2$O$_4$ in this case), and this beam, known as the probe beam, is focused onto the same region of the sample as the pump light. By monitoring the intensity of the reflected probe light with a photodetector, the transient reflectivity at the moments when the probe light pulses reach the sample surface can be monitored. By use of a variable optical delay line to change the relative arrival time of the pump and probe light pulses, one can track the transient optical reflectivity variation, dependent on the acoustic strain within the optical absorption depth of the probe, as a function of time. The temporal resolution is mainly determined by the optical pulse duration, and is less than 1 ps in this case. The longest delay time is determined by the length of the delay line, and is typically on the order of a few nanoseconds. It is also limited by the laser repetition period, i.e., 12.5 ns for an 80 MHz repetition rate.

The relative reflectivity variation typically obtained in picosecond laser acoustics is in the range 10^{-6} to 10^{-4}. To detect such small variations with sufficient signal-to-noise ratio, a lock-in detection method is used. An acousto-optic modulator (AOM), placed in the pump light path in Fig. 7.1, chops the pump light pulse train at about 1 MHz, and this results in a modulation of the probe light intensity at the same frequency. By use of a lock-in amplifier, a signal at the chopping frequency proportional to the tiny intensity changes induced in the probe beam can be extracted. An optical filter in front of the photodetector is used to eliminate pump light reaching it. Although the measurement applies to ultrafast phenomena in the picosecond regime, the required bandwidth for the photodetector is around the chopping frequency ($\sim$1 MHz in this case).

Acoustic waves can modulate the optical reflectivity through the photoelastic effect. This photoelastic modulation depends on the deformation of the medium, as determined by the propagating acoustic strain within the optical absorption depth of the probe light. Light propagation is governed by the permittivity of the medium, which is mainly determined by the electronic band structure. The reflection or scattering of light is caused by any inhomogeneity in the permittivity. When the acoustic strain modulates the atomic distance or configuration in the medium, the electronic band structure is modulated. In this way, the propagating acoustic strain modulates the permittivity. The permittivity variation in turn modulates the optical reflectivity, which is detected as intensity variations in the reflected probe light. This is basis of the photoelastic detection.

In fact the permittivity and its variation can be expressed by complex values in general, and both amplitude and phase modulations in the electric field of the reflected light can arise. In this article, the complex amplitude reflectivity variation for the electric field will be simply referred to as the optical reflectance variation. The term reflectivity variation will be reserved for intensity variations.

For partially transparent films, acoustic waves may modulate the optical reflectivity through surface and interface displacements of the sample surface. Such displacements alter the optical path and thus the phase of the reflected light. Interference between the light reflected from the surface of the sample and from buried interfaces will produce an intensity modulation of the probe light [3, 41, 44, 45].

Other mechanisms for probe reflectivity modulation have also been demonstrated, such as the combination of the piezoelectric effect and the electro-optic effect [65] or local rotations in the sample [27]. Other detection methods rely on the deflection of probe beam [18, 89].

7.2.2 Interferometric Setup

Any phase modulation in the reflected probe light involves the imaginary part of the optical reflectance variation. Since the light intensity only depends on the amplitude of the electric field, a simple intensity measurement does not reveal the optical phase modulation. To obtain both the real and imaginary parts of the optical reflectance variation, one should use an interferometer. Several setups have been proposed involving Mach–Zehnder interferometers [6], Sagnac interferometers [7, 9], or Michelson interferometers [8, 17, 90]. Figure 7.2 shows one such setup; the shaded area corresponds to a Sagnac interferometer with a common-path configuration. This allows one to measure the real part ρ and the imaginary part $\delta\phi$ of the relative reflectance variation $\delta r/r = \rho + i\delta\phi$ independently, where r is the optical amplitude reflectance and δr is the variation in r.

A brief description of the interferometer follows [7]. The probe light pulses fed to the interferometer are split into two beam paths using a polarizing beam splitter (PBS). Because of the difference between these two beam path lengths, the probe light pulses passing through these paths reach (the same position on) the sample surface with a time difference of several hundred picoseconds. The overall paths are chosen so that the probe pulses passing through the shorter path reach the sample before the pump light pulses arrive, whereas the probe pulses passing through the longer path reach the sample after the pump light pulses arrive. The probe light pulses reflected from the sample follow opposite paths owing to the PBS and a quarter wave plate (QWP) near the sample. With this configuration, the probe light pulses are unified again before heading to the photodetector. A polarizer (pol.) and another QWP in front of the photodetector are used to interfere probe light pulses that arrive at the sample before and after the pump light pulse arrival. In this way, the phase difference between the probe light pulses passing in the two paths is converted to an intensity variation. By scanning the delay line, the transient complex reflectance variation can be recorded as a function of the pump-probe delay time. Other parts of the setup are the same as the simpler one in Fig. 7.1.

7.2.3 Other Methods

There are various other related methods. The generation of (quasi) monochromatic acoustic waves can be achieved by the use of optical pulse trains [12, 52, 62, 91]. Another method involves laser-induced gratings to generate and detect SAWs (see Sect. 7.4) with a given wave number [29, 92]. Surface plasmons can also be used for the detection [93, 94]. The use of parallel-processing in the detection has been suggested to reduce the measurement time [95].

7.3 Excitation and Detection of Bulk Acoustic Waves

Both bulk and SAWs are generated by irradiation with light pulses. Picosecond laser ultrasonics is concerned with the observation of bulk acoustic waves, typically within a few microns of the sample surface. This depth is usually smaller than the pump spot diameter. In this case, and with a flat sample surface, bulk acoustic waves propagating perpendicular to the surface are mainly generated and detected.

As an example, a multilayer sample may have bulk acoustic waves generated near the sample surface. Part of the acoustic field propagating in the depth direction may be reflected at each interface and return to the sample surface. For typical longitudinal sound velocities in solids on the order of several $km\ s^{-1}$ (or $nm\ ps^{-1}$), the round trip time in a film of thickness 100 nm is several tens of picoseconds. A transient optical reflectance measurement for this sample with the setup described in Sect. 7.2 allows the evaluation of the round trip time of the acoustic waves and, in turn, allows one to extract information on the geometry of the multilayer structure with nanometer precision [6, 40, 44].

More detailed analyses of such experimental results can reveal various physical properties, such as elastic properties [2, 3, 8, 14, 22, 24, 26, 35, 40, 43, 46, 47, 54], ultrasonic absorption [2, 39, 42, 58, 52], photoelastic properties [3, 6, 11, 17, 20, 27, 32, 40, 44, 59, 68], piezoelectric properties [30, 38, 72], thermal diffusion [2, 8, 17, 29], or ultrafast dynamics of photo-excited carriers [2, 4, 5, 13, 17, 27, 31, 32,

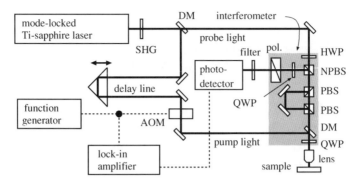

Fig. 7.2 Schematic diagram of the typical setup for picosecond laser ultrasonics measurements with an interferometer (*shaded region*), in this case based on a Sagnac design. *SHG* second harmonic generation crystal, *AOM* acousto-optic modulator, *HWP* half-wave plate, *QWP* quarter wave plate, *pol.* polarizer, *PBS* polarizing beam splitter, *NPBS* nonpolarizing beam splitter, *DM* dichroic mirror

38]. Some examples concerning metals and semiconductors, including anisotropic media, will be presented below.

7.3.1 Thin Metal Films

Consider the absorption of ultrashort light pulses at a metal surface at room temperature. A typical optical penetration depth in metals for visible light is ∼10 nm. Electrons are excited within this depth. The absorbed energy per electron can be up to several eV, which is much greater than the thermal equilibrium energy of an electron at room temperature (∼0.025 eV). This excess energy becomes redistributed among the whole system of electrons and lattice phonons. Since the phonons interact with the electrons less efficiently than the electrons interact with each other, the initial interactions occur mostly within the electron system. In picosecond laser ultrasonics measurements, the temperature rise in the electron system is typically on the order of several 100 K. The excess energy distributed in the electron system is eventually transferred to the phonon system via the electron–phonon interaction. After a few picoseconds, the whole system is thermalized to a new temperature a few K above the initial state. In metals, the temperature rise in the phonon system produces a thermal stress distribution which in turn launches the acoustic waves (as coherent acoustic phonons).

The thermalization process between the electron and phonon systems involves the diffusion of excited electrons. Because of this, the region of elevated phonon temperature tends to be broader than the region of the initially excited electrons (≃ the optical penetration depth). The spatial width of the generated acoustic pulse may be roughly approximated as the depth of this hot phonon region. More detailed quantitative discussion can be made using a two-temperature model (TTM) [96]. The model assumes that the electron and phonon systems are thermalized almost independently, that energy transfer between them occurs via the electron–phonon interaction, and that the electrons diffuse during the thermalization process. The diffusion in the phonon system is less efficient than that in the electron system, and can be neglected. The acoustic wave generation and propagation is described by the elastic wave equation with the source term determined by the lattice temperature rise based on the TTM.

In metals such as Au, Ag, and Al, having relatively large thermal conductivities and small electron–phonon interactions, the excited electrons diffuse ~50–100 nm before they thermalize with the phonon system. On the other hand, in metals such as Cr and Ni, having relatively small thermal conductivities and large electron–phonon interactions, the diffusion length of the excited electrons is comparable to the optical penetration depth ~10 nm. So the frequency spectrum of the generated acoustic waves for the former metals extends to ~10 GHz, whereas that for the latter metals extends to ~100 GHz.

Figure 7.3 shows typical experimental results for a Cr film of thickness 190 nm made by electron beam deposition on a Si (100) substrate [13]. Measurements were made using the setup shown in Fig. 7.2. The real and imaginary parts of the relative reflectance variation are plotted as a function of delay time. The sharp peak near 0 ps in Fig. 7.3a is caused by the initial excitation of the electron system. Small peaks are observed with a period of ~60 ps on a slowly decaying background signal. The background is due to the temperature variation caused by three-dimensional thermal diffusion, whereas the small peaks are due to the periodic arrival of acoustic pulses bouncing back and forth inside the Cr film.

Figure 7.3b shows a magnified view of the first echo. The solid lines show the experimental data. The dashed lines are based on a model involving thermoelastic

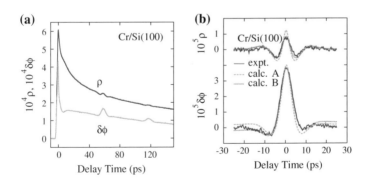

Fig. 7.3 **a** Transient reflectance change of a Cr/Si(100) sample. ρ and $\delta\phi$ are the real and imaginary parts of the relative reflectance variation. **b** Magnified view of the reflectance variation caused by the first echo. The origin of the temporal axis is shifted so that it corresponds to the center of the pulse. *Solid line* (expt., *red*): experimental result, *dashed line* (calc. A, *green*): simulation ignoring diffusion processes, and *dotted line* (calc. B, *blue*): simulation including electron diffusion (two-temperature model)

stress that arises from a step-like temperature rise and with a spatial extent determined by the optical penetration depth only. The dotted lines are based on the TTM which takes account of electron diffusion and the electron–phonon interaction. In both cases, the acoustic wave generation and propagation are calculated from the spatiotemporal lattice temperature field. The relative optical reflectance variation is calculated by considering both the inhomogeneous modulation of the permittivity through the photoelastic effect and the surface displacement caused by the acoustic wave propagation. The echo shape calculated without diffusion processes (dashed line) is symmetric, a result of a strain pulse that is anti-symmetric in shape being reflected from the free surface with a flipped polarity (corresponding to a π phase change). In contrast, the echo shape calculated from the TTM (dotted line) is asymmetric, a result of an asymmetric strain pulse shape being so reflected. The latter echo is also broader than the former. The TTM model gives better agreement with experiment than the simpler thermoelastic model ignoring diffusion processes. In this way, theoretical models for the ultrafast relaxation of the electron and phonon systems can be tested with picosecond laser ultrasonics, and values for the electron–phonon interaction coefficient and photoelastic constants can be derived.

7.3.2 Semiconductor Quantum Wells

The mechanism for acoustic wave generation by ultrashort light pulse irradiation in metals is based on thermoelastic stress (thermal expansion). In semiconductors, another mechanism based on the deformation potential also plays an important role. The deformation potential is the variation of the conduction and valence band energy due to the strain. The inverse process involves the excitation of electrons to the conduction band or holes to the valence band, thus producing a stress distribution that depends on the excited electron/hole density.

Picosecond laser ultrasonics measurements were carried out on a GaAs/$Al_{0.3}Ga_{0.7}As$ quantum well (QW) sample formed on a GaAs (100) substrate by metallorganic vapor phase epitaxy [68]. The sample has three wells with different well widths (see Fig. 7.4a). The optical setup is similar to Fig. 7.2, but involving the use of two synchronized mode-locked Ti:sapphire lasers for the pump and probe light beams with independently-tunable wavelengths. The probe light wavelength is fixed at 415 nm, which has a penetration depth $\sim$10 nm for the QW sample. Only the near-surface region of the sample is therefore involved in the optical detection.

Figure 7.4b–d shows the real and imaginary parts of the relative reflectance variation as a function of delay time at various pump photon energies. The $Al_{0.3}Ga_{0.7}As$ layer has a bandgap at 1.83 eV, and is transparent for all the pump photon energies used in the experiment (1.47–1.64 eV). Because of the quantum confinement effect, the transition energy for the quantum well increases on decreasing the well width. The pump photon energies 1.47, 1.50, and 1.64 eV used are just above the lowest transition energy for QW2, QW1, and QW3, respectively. The pump light is absorbed at the wells with a transition energy less than the pump photon energy, as well as in

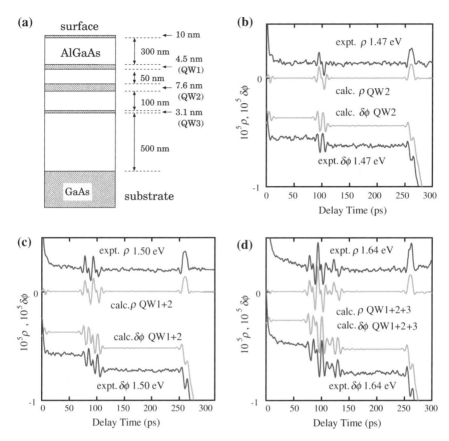

Fig. 7.4 a Structure of a GaAs/Al$_{0.3}$Ga$_{0.7}$As quantum well sample. **b–d** Experimental results (expt.) and simulation (calc.) for the quantum well sample with various pump photon energies; **b** 1.47 eV, **c** 1.50 eV, and **d** 1.64 eV. ρ and $\delta\phi$ are the real and imaginary parts of the relative reflectance variation

the substrate. The decrease in $\delta\phi$ after 250 ps is attributed to the arrival at the sample surface of a compressive acoustic pulse generated in the substrate. In this case, the surface is displaced in the outward direction. The wiggles in $\delta\phi$ and ρ at ~100 ps are attributed to the arrival at the sample surface of the acoustic pulses generated in the QW layers. The shape of these signals depends on the unipolar strain pulse shape and the flipping of the strain pulse on reflection from the free surface. On increasing the photon energy, the duration of the wiggles increases. This can be easily understood since the higher energy photons excite more QWs than the lower energy photons do.

A simulation was also carried out, as shown in Fig. 7.4b–d. Generation of acoustic waves in the QWs is based on the deformation potential mechanism, the propagation of the acoustic waves is based on the acoustic wave equations, and the detection of the acoustic waves is based on light scattering theory for a medium with inhomogeneously modulated permittivity. The excited electrons and holes are confined in the QWs, and so it is not necessary to consider carrier diffusion. Although the spatial width of the generated acoustic pulses is expected to be comparable to the QW width, the wiggling period is much longer than that expected for acoustic pulses of the QW width. This discrepancy arises because the probe light effectively senses the strain field within the optical penetration depth rather than just at the sample surface. The simulation allows the quantitative evaluation of the photoelastic constants for GaAs and $Al_{0.3}Ga_{0.7}As$, the ultrasonic absorption coefficient of $Al_{0.3}Ga_{0.7}As$, and the effective extinction coefficient for the QW layers.

In addition to evaluating the physical properties of the sample, the acoustic wave generation in QWs can be used to realize ultrahigh-frequency phonon transducers. Considering the value of the longitudinal sound velocity in GaAs ($\sim 5\,km\,s^{-1}$), acoustic pulses with nm spatial extent should have frequency components up to at least 1 THz. The Fourier spectrum of the acoustic pulses detected in the above experiment is limited to 0.5 THz mainly because of frequency-dependent ultrasonic absorption. Acoustic pulses with frequency components above 1 THz have been reported in multiple quantum wells [30, 63, 66, 72].

7.3.3 Shear Acoustic Waves

In general, acoustic waves propagating in a specific direction in a solid exhibit three different polarizations. In an isotropic medium, for example, there is one longitudinal mode in which the displacement is parallel to the wave vector, and two transverse (shear) modes in which the displacement is perpendicular to the wave vector. For the two transverse modes, the polarization vectors are orthogonal and the sound velocities are degenerate. The propagation of the longitudinal mode is accompanied by a microscopic volume variation, whereas the propagation of the transverse mode is accompanied by a shear deformation. In the samples considered in Sects. 7.3.1 and 7.3.2, the acoustic waves generated by light absorption that travel directly away from the surface are limited to the longitudinal mode owing to the symmetry of the system. It is, however, desirable to generate and detect shear waves for better access to the elastic properties of the medium.

To generate shear acoustic waves in picosecond laser ultrasonics, one should break the symmetry of the sample or the measurement configuration. This has been achieved using acoustic mode conversion at an interface between an anisotropic medium and an isotropic medium [10, 20, 25], by the use of anisotropic stress generation such as by the thermal expansion mechanism in an anisotropic medium [14, 25, 27, 28, 37, 69], and by the use of acoustic waves propagating along a direction not perpendicular to the sample surface [16, 23]. Optically generated high-frequency shear acoustic waves have also been used to study the high-frequency viscoelastic properties of liquids [52].

Shear acoustic wave generation and detection was studied using a hexagonal Zn single crystal [14]. It is well known that Zn has a large anisotropy [97]. The sample was prepared by cutting the Zn single crystal with its c-axis having an angle of 45° to

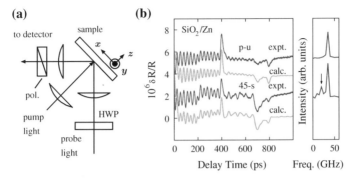

Fig. 7.5 **a** Schematic diagram of the experimental setup for shear wave generation and detection. *HWP* half-wave plate, *pol.* polarizer. **b** Transient relative reflectivity variation of a SiO_2/Zn sample. *Left hand side* from *top* to *bottom*; experimental result for *p-u* configuration (expt.), simulation for *p-u* configuration (calc.), experimental result for 45-*s* configuration (expt.), simulation for 45-*s* configuration (calc.). *Right hand side* Fourier transform of the experimental curves on the *left hand side*. The *arrow* shows the frequency component associated with the shear acoustic waves in SiO_2

the sample surface, and by depositing a SiO_2 film of a thickness $\sim 1\,\mu m$ on the sample surface by rf (radio-frequency) sputtering. Figure 7.5a shows a schematic diagram of the experimental setup. The x, y, and z coordinate axes are also defined in this figure. The pump and probe light wavelengths are 814 and 407 nm, respectively. The probe light beam is obliquely incident on the sample surface with an incident angle of 45°. To control the probe light polarization, a half-wave plate and a polarizer are placed on the incident and reflected sides of the probe beam path, respectively. The c-axis of the sample is in the yz plane.

Figure 7.5b shows the relative reflectivity variation as a function of delay time for two different polarization configurations. The notation *p-u* indicates that the incident probe light is *p*-polarized and that no analyzer (polarizer) is used for the reflected light. The notation 45-*s* indicates that the incident probe light is linearly polarized with its polarization plane at 45° to the incident plane, and the *s*-polarized component of the reflected light is observed. Either case results in a relatively narrow peak, a first echo, at ~ 400 ps being recorded. This is attributed to a longitudinal acoustic pulse, that is generated at the SiO_2/Zn interface, propagating into the SiO_2 layer, being reflected at the top surface, and then re-entering the Zn layer. The reflectivity variation is caused by the modulation of the permittivity in the Zn substrate through the photoelastic effect. Part of the returning longitudinal pulse is again reflected back to the SiO_2 layer and gives rise to a second echo at 800 ps when it re-enters the Zn layer. The polarity of the 800 ps peak is flipped with respect to the 400 ps peak. This can be understood by considering the acoustic impedance of the layers: the polarity of the acoustic strain is flipped by reflection at the top surface, whereas it is not flipped on reflection at the SiO_2/Zn interface. A relatively broad peak at ~ 700 ps in either polarization configuration is also noticeable. This is attributed to the acoustic pulse, generated at the SiO_2/Zn interface and propagating to the SiO_2 layer as shear acoustic waves, being reflected at the top surface and then re-entering the Zn layer. The shape of these echoes is asymmetric owing to the strain pulse shape and the region of detection: both the longitudinal and shear strain pulses generated in the Zn layer that propagate to the SiO_2 layer are unipolar, and the echo detection relies on the photoelastic effect in the buried Zn substrate where the strain pulse propagation is unidirectional.

Another important feature of these results is an oscillation mainly observed between 0 and 400 ps for either polarization configuration. This is known as a Brillouin oscillation, which is caused by the interference of probe light reflected from the surface or interface with that reflected from the moving acoustic pulse. The Brillouin oscillation frequency is proportional to the sound velocity. For the p-u configuration, the oscillation amplitude is almost constant, whereas the amplitude is periodically modulated for the 45-s configuration. These results indicate that the former oscillation consists of a single frequency component, whereas the latter consists of multiple frequency components. The right-hand side of Fig. 7.5b shows the Fourier spectrum of the relative reflectivity variation between 0 and 400 ps. As expected, the p-u configuration has a single peak at 35 GHz, whereas the 45-s configuration has two peaks at 35 and 20 GHz. From a knowledge of the longitudinal and shear sound velocities in glassy SiO_2, the 35 and 20 GHz peaks can be attributed to longitudinal and shear acoustic waves, respectively, in the SiO_2 layer.

For the orientation of the crystal axes and for the excitation conditions in this experiment, it is expected that the shear waves in the SiO_2 layer yield a finite yz strain component that modulates the yz permittivity component through the photoelastic effect. To observe this permittivity modulation as a reflectivity variation, the probe light electric field must have an finite z component in the medium. This is why an oblique probe light incidence setup is used. In addition, according to detailed theoretical considerations, the modulation in the yz permittivity component scatters the p-polarized incident probe light to s-polarized reflected light, and vice versa. If one used pure p-polarized light for the incident probe light, the electric field amplitude of the reflected s-polarized light would be proportional to strain, and thus the reflected s-polarized light intensity would be quadratic in the strain. For the typical strain amplitudes in picosecond laser ultrasonics, the expected reflected s-polarized light intensity would in that case be too small to be detected. To overcome this difficulty, we use a mixture of s- and p-polarized light for the incident probe light, and exploit the interference between the s component scattered from the p component by the yz permittivity modulation and the s component originally involved in the non-perturbed reflected light, so that the reflected light intensity variation is proportional to the strain. This is why a 45-s configuration for the shear wave detection is used. It also explains why the p-u configuration does not show a Brillouin oscillation arising from shear waves.

For a quantitative analysis of the experimental results, a model consisting of the generation, propagation, and detection of the acoustic waves is used [27]. The acoustic wave generation is treated using the TTM described in Sect. 7.3.1. The anisotropic acoustic wave equation and anisotropic thermal expansion is also taken into account. For the propagation, the anisotropy of the medium, the acoustic mode conversion at the SiO_2/Zn interface, and the ultrasonic absorption are considered. For the detection, a light scattering theory for the probe light obliquely incident on an anisotropic multilayer with inhomogeneously modulated permittivity is constructed, and this is used to calculate the reflectivity variation caused by the acoustic wave propagation. The simulation results are also plotted in Fig. 7.5, showing good agreement with experiment. The photoelastic tensor components for Zn and SiO_2, the ultrasonic absorption coefficient in SiO_2, the electron–phonon coupling constant for Zn are quantitatively retrieved in this simulation.

7.4 Excitation and Imaging of Surface Acoustic Waves

The surface of a medium supports SAWs whose elastic energy is mostly confined within a depth on the order of the acoustic wavelength. SAWs are important for device applications. SAW filters, for example, are widely used in GHz communication equipment, such as mobile phones. To understand SAWs, it is important to be able to image their propagation [98–120].

In the experimental setup for picosecond laser ultrasonics such as the one shown in Fig. 7.2, the pump light pulses also generate SAWs. One can obtain time-resolved two-dimensional images of SAW propagation by controlling the spatial positions of the pump and probe light spots independently. The frequency range of SAWs obtained with this method is typically limited to several GHz, and usually the term picosecond laser ultrasonics does not include SAW imaging. However, the experimental techniques have much in common.

When two independent lenses are used for the pump and probe light focusing on the sample, it is possible to scan the probe light spot position on the sample just by scanning the probe focusing lens in the plane perpendicular to its optical axis. This sort of optical setup is conveniently used for imaging samples consisting of opaque films on transparent substrates, since in this case the pump and probe light pulses can be focused onto the film from opposite sides [105]. In the case of opaque substrates, however, the pump and probe light should be focused from the same side, and it may be difficult to use two independent lenses for the pump and probe light because of the insufficient space to place two lenses [121].

To solve this problem, an optical scanning system was developed using a combination of a biaxial rotating mirror and a pair of lenses placed in a 4f configuration [90]. The setup is shown schematically in Fig. 7.6. The pump and probe light beams pass through a common microscope objective. By tilting a biaxial rotating mirror, the probe light incidence angle on the microscope objective is altered, but the beam position stays at the entrance aperture of the objective owing to the $4f$ system. The objective converts this variation of the incident angle to a variation in the lateral spot position on the sample surface. With this method it is possible to obtain a two-dimensional image of the reflectivity variation across an area up to $\sim 500 \times 500\,\mu m^2$. Combining the scanning system with the time-resolved reflectivity measurement system of Fig. 7.2, the spatiotemporal evolution of the propagating SAW field can be

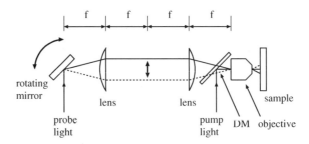

Fig. 7.6 Schematic diagram of the two-dimensional spatial scanner for the probe light in SAW imaging. By rotating the mirror, the focusing spot moves across the sample surface. f denotes the focal length of the lenses. *DM* dichroic mirror

obtained by recording multiple two-dimensional images at varying pump-probe delay times. A 4-m long delay line is used in order to cover one period of the laser pulse repetition (~13 ns).

The minimum diameter of the pump and probe light spots is limited by optical diffraction to ~1 μm. The shortest SAW wavelength which can be generated by the absorption of optical pulses is therefore limited by the optical spot size. In a typical solid, this SAW wavelength for a ~1 μm optical spot corresponds to a frequency of ~1 GHz. As described in Sect. 7.2, the interferometer detects the phase difference between consecutive probe light pulses that arrive at the sample at an interval of a few hundred picoseconds. So for frequencies in the 100 MHz to the GHz region, this phase difference can be regarded as being proportional to the surface velocity normal to the surface.

In the following sections, we review results obtained using this SAW imaging technique for glasses, crystals, and phononic crystals.

7.4.1 Glasses and Crystals

Figure 7.7a shows an image of SAW propagation for a crown glass plate of thickness 1 mm coated with a gold film of thickness 70 nm [105]. The imaged area is 200 × 200 μm². The pump light pulses are focused to the point at the center of the image, and the normal surface velocity 7.0 ns after the pump pulse arrival at the sample surface is plotted in a color scale.

In this case, both the substrate and film are isotropic in the lateral direction. This is consistent with the observed ring-shaped axisymmetric wave packets emerging from the excitation point. Within the observed area, wave packets generated within ~4 periods of the laser repetition are involved as a series of sets of concentric rings. This evident dispersion is caused by the fact that the sound velocity for the gold film is slower than that for the crown glass plate; the waves with shorter wavelength travel within a shallower depth, and are more affected by the slower sound velocity

(a) **(b)**

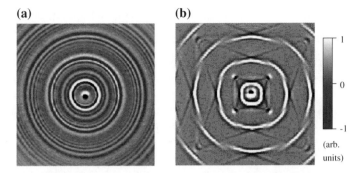

Fig. 7.7 Snapshot of the surface displacement velocity field caused by SAW propagation. **a** 200 × 200 μm^2 area of a Au/crown glass sample at a moment 7.0 ns after the pump pulse arrival. **b** 150 × 150 μm^2 area of a Au/TeO$_2$ (001) sample at a moment 4.9 ns after the pump pulse arrival. White corresponds to a surface velocity directed outward from the sample surface

of the gold film. From such observations it is possible to obtain information on the structure of the sample in the depth direction [121–125].

Anisotropy in the medium gives rise to striking effects in the acoustic wave propagation. Figure 7.7b shows an image of SAW propagation for a TeO$_2$ (001) substrate coated with a gold film of thickness 60 nm [105]. The imaged area is 150 × 150 μm^2. The pump light pulses are focused to a point at the center of the image, and the normal surface velocity at 4.9 ns after the pump light pulse arrival at the sample surface is plotted in a color scale. The large elastic anisotropy of tetragonal TeO$_2$ is evident in this SAW image.

The complicated pattern shown in Fig. 7.7b indicates that the group velocity varies with the propagation direction; it clearly shows the phonon focusing effect in which the acoustic wave energy is focused to certain directions [98, 126–129]. A series of images at different delay times (not shown here) can be used to obtain images in ω-k space (related to the acoustic dispersion relation) by taking Fourier transforms [107, 112].

These experimental results for wavefront shape all agree well with theoretical calculations [130]. The results shown here are for uniform samples, but the experimental technique can be applied to more complicated structures, such as polycrystalline samples with grains larger than several μm in size, in order to elucidate hidden sub-surface structure [114].

7.4.2 Phononic Crystals

Control of acoustic wave propagation may be achieved by artificial periodic structures which consist of two or more media with different acoustic impedances. Such structures are called phononic crystals. Just as band gaps in the electronic structure of semiconductors arise from the periodic arrangement of the atoms, specially designed phononic crystals show a phononic bandgap in which the acoustic waves within certain frequency ranges cannot propagate [131–135].

Phononic crystals are classified as being either one- two- or three-dimensional depending on the spatial dimension of the periodic structure. Here, we discuss SAW imaging on a one-dimensional (1D) phononic crystal [115]. The sample consists of periodic stripes of Cu and silicon oxide formed on a Si (100) substrate. Each stripe has a thickness of 800 nm and a width of 2 μm. Figure 7.8a shows a cross section of the sample. To enhance the excitation efficiency of the acoustic waves, a gold film of thickness 30 nm is deposited on the top surface.

Using the method described in Sect. 7.4, time-resolved 2D images of the SAWs are obtained. By 2D spatial and temporal Fourier transforms of the spatiotemporal data, the Fourier amplitude can be obtained in 2D k-space at each frequency. Figure 7.8b shows an example of a Fourier amplitude image at 534 MHz: the modulus of the Fourier amplitude is represented with a color scale over the (k_x, k_y) wavenumber plane. The periodicity of the sample is along the k_x direction. A finite value of the Fourier amplitude is observed when the combination of the wave vector k and the angular frequency ω satisfies the dispersion relation of the corresponding acoustic mode.

According to the spatial period of 4 μm along the x direction, the Brillouin zone boundaries are defined by $k_x = \pm 0.79 \mu m^{-1}$ (indicated by the arrows in the figure). The Fourier amplitude image appears folded at the Brillouin zone boundary because the dispersion relation exhibits translational symmetry defined by the reciprocal lattice vector. The red curves are discontinued in the region of the zone boundary. This indicates that there is no acoustic mode propagating in the x direction at 534 MHz. In other words, the sample has a directional phononic bandgap.

Time-resolved SAW imaging is also appropriate for investigating the acoustic properties of 2D phononic crystals [117, 119].

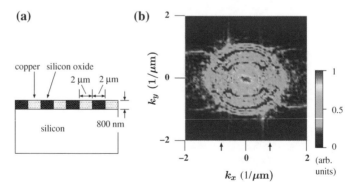

Fig. 7.8 **a** Cross section of a 1D phononic crystal sample. **b** The modulus of the Fourier amplitude at 534 MHz is plotted in k-space. The *small arrows* indicate the first Brillouin zone boundary, arising from the periodicity of the sample

7.4.3 SAW Imaging Through the Photoelastic Effect

In the above examples of SAW imaging, the interferometer is used to detect the velocity of the out-of-plane motion of the surface. However, imaging is also possible by detecting the reflectivity variation caused by the photoelastic effect. In some cases, measurements based on the photoelastic effect can bring additional information owing to the higher degree of freedom allowed for the optical configuration, such as in the choice of the probe light polarization. This is especially so if there are some modes which have no normal surface displacement, and so cannot be detected by an interferometer. Imaging through the photoelastic effect allows the detection of such modes by a suitable choice of the optical polarization configuration.

Measurements were reported for a crown glass plate of 1 mm thickness coated with a gold film of 45 nm thickness [120]. The optical setup is similar to that in Fig. 7.1 combined with the scanning system in Fig. 7.6. The setup around the sample is modified to allow for various optical polarization configurations and for pump and probe light incidence from opposite sides of the sample, as shown in Fig. 7.9. The incident probe light polarization can be chosen as left or right circularly polarized, and the Y (perpendicular to the plane of the page) linear polarization component of the reflected light is directed onto the photodetector. The notation R-Y and L-Y to specify the above-mentioned polarization configurations will be used in the description below.

Figure 7.10a and b shows images of the SAW propagation for the R-Y and L-Y configurations, respectively. The imaging area is $140 \times 140 \, \mu m^2$. The pump light pulses are focused to a point at the center of the image, and the intensity variation ΔI of the reflected probe light at 11.7 ns after the moment of pump light arrival at the sample surface is plotted in a color scale. The vertical direction in the image corresponds to the Y direction. For later use, we define the x and y axes as the horizontal and vertical directions in the image, i.e., in the sample plane. Each image

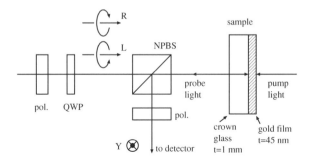

Fig. 7.9 Schematic diagram of the optical setup for imaging SAWs through the photoelastic effect with a Au/crown glass sample. The pump light is incident from the film side whereas the probe light is incident from the substrate side. *R* right circular polarization, *L* left circular polarization, *Y* linear polarization perpendicular to the plane of the page. *QWP* quarter wave plate, *NPBS* nonpolarizing beam splitter, *pol.* polarizer. *t* refers to the thickness

(a) $\Delta I_{\text{R-Y}}$ **(b)** $\Delta I_{\text{L-Y}}$ **(c)** $\Delta I_{\text{L-Y}} - \Delta I_{\text{R-Y}}$

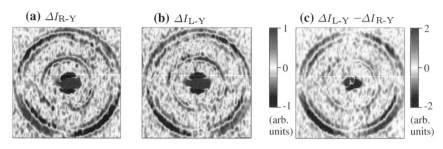

Fig. 7.10 Snapshot of an intensity variation image caused by SAW propagation in a Au/crown glass sample. **a** $\Delta I_{\text{R-Y}}$: using right circularly polarized probe light and detection of the Y linear polarization component. **b** $\Delta I_{\text{L-Y}}$: using left circularly polarized probe light and detection of the Y linear polarization component. A $140 \times 140\,\mu\text{m}^2$ area imaged at the moment 11.7 ns after the pump pulse arrival. **c** $\Delta I_{\text{L-Y}} - \Delta I_{\text{R-Y}}$: a subtraction image which shows a $\sin 2\theta$ dependence, where θ is the angle between the propagation direction and the horizontal axis

contains two rings: the inner one corresponds to the Rayleigh-like wave and the outer one to the Sezawa wave. Each ring shows a significant intensity and signal-polarity variation depending on the propagation direction and the polarization configuration.

A detailed theoretical analysis that considers the photoelastic effect and the light scattering by the inhomogeneously modulated permittivity was also presented [120]. The theory predicts that a subtraction image $\Delta I_{\text{L-Y}} - \Delta I_{\text{R-Y}}$ should depend solely on the in-plane displacement or the xy strain component η_{xy}. Figure 7.10c shows an experimental subtraction image which exhibits a $\sin 2\theta$ intensity dependence, where θ is an angle of the propagation direction with respect to the x axis. This arises because the strain component η_{xy} for a circularly shaped wave packet on a laterally isotropic medium has a $\sin 2\theta$ dependence. This result indicates that SAW imaging through the photoelastic effect can reveal in-plane displacements not detectable by the interferometric method. The dual use of the photoelastic and interferometric methods would allow more complete information on the nature of propagating SAWs to be extracted than a measurement using either method alone.

7.5 Summary

We have presented an overview of optical methods to generate and detect acoustic waves up to terahertz frequencies in various media and then given examples taken from our own work.

In Sect. 7.2, we presented some typical experimental setups for picosecond laser ultrasonics. These methods are based on the optical pump-probe technique. The pump light pulses generate acoustic waves in a sample and the probe light pulses are used to detect the transient optical reflectivity variation and/or complex reflectance variation caused by the acoustic wave propagation. Measurement of the real and imaginary parts of the complex reflectance variations can be achieved by the use of interferometry.

In Sect. 7.3, we dealt with bulk acoustic waves. For metal thin-film samples, we described how longitudinal acoustic waves could be generated thermoelastically. The optically excited electrons diffuse significantly before they thermalize and produce a thermal stress, thus reducing the excited acoustic frequencies to the sub-100 GHz range. For GaAs/AlGaAs quantum well samples, we showed how near-THz longitudinal acoustic waves could be detected at the sample surface and how varying the pump photon energy could be used to achieve selective acoustic generation. The acoustic wave generation is mainly governed by the deformation potential in this case. In addition, we discussed how shear acoustic waves could be generated by anisotropic thermal expansion in crystals.

In Sect. 7.4, we reviewed experimental methods for measuring the transient surface velocity normal to the surface and producing two-dimensional images of the SAW propagation at GHz frequencies. Applications to uniform isotropic and anisotropic media, and to a one-dimensional phononic crystal, were reviewed. We showed how spatiotemporal data can be Fourier transformed to obtain two-dimensional images in k-space related to the acoustic dispersion relation. We also explained how imaging with the photoelastic effect is useful for obtaining images related to in-plane strain not accessible by optical interferometry.

In conclusion, ultrafast optical methods for generating and detecting acoustic waves can be used to determine the geometry of structures as well as various physical properties of the media that comprise them. These methods are particularly promising for the investigation of nanoscale and microscale structures. In particular, structures based on metamaterials, such as phononic crystals, can be investigated. By accessing various acoustic polarizations and modes, such as longitudinal or shear polarizations and surface, plate or interface modes, a wider range of sample properties can be investigated.

References

1. C. Thomsen, J. Strait, Z. Vardeny, H.J. Maris, J. Tauc, J.J. Hauser, Phys. Rev. Lett. **53**, 989 (1984)
2. C. Thomsen, H.T. Grahn, H.J. Maris, J. Tauc, Phys. Rev. B **34**, 4129 (1986)
3. O.B. Wright, J. Appl. Phys. **71**, 1617 (1992)
4. G. Tas, H.J. Maris, Phys. Rev. B **49**, 15046 (1994)
5. O.B. Wright, Phys. Rev. B **49**, 9985 (1994)
6. B. Perrin, B. Bonello, J.C. Jeannet, E. Romatet, Prog. Nat. Sci. **S6**, S444 (1996)
7. D.H. Hurley, O.B. Wright, Opt. Lett. **24**, 1305 (1999)
8. C.J.K. Richardson, M.J. Ehrlich, J.W. Wagner, J. Opt. Soc. Am. B **16**, 1007 (1999)
9. M. Nikoonahad, S. Lee, H. Wang, Appl. Phys. Lett. **76**, 514 (2000)
10. D.H. Hurley, O.B. Wright, O. Matsuda, V.E. Gusev, O.V. Kolosov, Ultrasonics **38**, 470 (2000)
11. A. Devos, C. Lerouge, Phys. Rev. Lett. **86**, 2669 (2001)
12. Y. Matsuda, C.J.K. Richardson, J.B. Spicer, IEEE Trans. Ultrason. Ferroelectr. Freq. Control **49**, 915 (2002)

13. T. Saito, O. Matsuda, O.B. Wright, Phys. Rev. B **67**, 205421 (2003)
14. O. Matsuda, O.B. Wright, D.H. Hurley, V.E. Gusev, K. Shimizu, Phys. Rev. Lett. **93**, 095501 (2004)
15. H. Hébert, F. Vidal, F. Martin, J.C. Kieffer, A. Nadeau, T.W. Johnston, A. Blouin, A. Moreau, J.P. Monchalin, J. Appl. Phys. **98**, 033104 (2005)
16. C. Rossignol, J.M. Rampnoux, M. Perton, B. Audoin, S. Dilhaire, Phys. Rev. Lett. **94**, 166106 (2005)
17. T. Dehoux, M. Perton, N. Chigarev, C. Rossignol, J.M. Rampnoux, B. Audoin, J. Appl. Phys. **100**, 064318 (2006)
18. N. Chigarev, C. Rossignol, B. Audoin, Rev. Sci. Instrum. **77**, 114901 (2006)
19. J. Vollmann, D.M. Profunser, J. Bryner, J. Dual, Ultrasonics **44**, e1215 (2006)
20. T. Pezeril, N. Chigarev, P. Ruello, S. Gougeon, D. Mounier, J.M. Breteau, P. Picart, V. Gusev, Phys. Rev. B **73**, 132301 (2006)
21. J. Bryner, D.M. Profunser, J. Vollmann, E. Mueller, J. Dual, Ultrasonics **44**, e1269 (2006)
22. B. Audoin, M. Perton, N. chigarev, C. Rossignol, J. Phys. Conf. Ser. **92**, 012028 (2007)
23. T. Dehoux, N. Chigarev, C. Rossignol, B. Audoin, Phys. Rev. B **76**, 024311 (2007)
24. H. Ogi, M. Fujii, N. Nakamura, T. Yasui, M. Hirao, Phys. Rev. Lett. **98**, 195503 (2007)
25. T. Pezeril, P. Ruello, S. Gougeon, N. Chigarev, D. Mounier, J.M. Breteau, P. Picart, V. Gusev, Phys. Rev. B **75**, 174307 (2007)
26. N. Nakamura, H. Ogi, T. Shagawa, M. Hirao, Appl. Phys. Lett. **92**, 141901 (2008)
27. O. Matsuda, O.B. Wright, D.H. Hurley, V. Gusev, K. Shimizu, Phys. Rev. B **77**, 224110 (2008)
28. D. Mounier, E. Morozov, P. Ruello, J.M. Breteau, P. Picart, V. Gusev, Eur. Phys. J. Special Topics **153**, 243 (2008)
29. A. Harata, T. Sawada, Jpn. J. Appl. Phys. **32**, 2188 (1993)
30. C.K. Sun, J.C. Liang, X.Y. Yu, Phys. Rev. Lett. **84**, 179 (2000)
31. N.V. Chigarev, D.Y. Paraschuk, X.Y. Pan, V.E. Gusev, Phys. Rev. B **61**, 15837 (2000)
32. O.B. Wright, B. Perrin, O. Matsuda, V.E. Gusev, Phys. Rev. B **64**, 081202(R) (2001)
33. D.H. Hurley, K.L. Telschow, Phys. Rev. B **66**, 153301 (2002)
34. N.C.R. Holme, B.C. Daly, M.T. Myaing, T.B. Norris, Appl. Phys. Lett. **83**, 392 (2003)
35. R. Côte, A. Devos, Rev. Sci. Instrum. **76**, 053906 (2005)
36. S. Wu, P. Geiser, J. Jun, J. Karpinski, J.R. Park, R. Sobolewski, Appl. Phys. Lett. **88**, 041917 (2006)
37. Y.C. Wen, T.S. Ko, T.C. Lu, H.C. Kuo, J.I. Chyi, C.K. Sun, Phys. Rev. B **80**, 195201 (2009)
38. P. Babilotte, P. Ruello, G. Vaudel, T. Pezeril, D. Mounier, J.M. Breteau, V. Gusev, Appl. Phys. Lett. **97**, 174103 (2010)
39. H.N. Lin, R.J. Stoner, H.J. Maris, J. Tauc, J. Appl. Phys. **69**, 3816 (1991)
40. O.B. Wright, Opt. Lett. **20**, 632 (1995)
41. V.E. Gusev, Acustica Acta Acust. **82**, S37 (1996)
42. C.J. Morath, H.J. Maris, Phys. Rev. B **54**, 203 (1996)
43. Y.C. Lee, K.C. Bretz, F.W. Wise, W. Sachse, Appl. Phys. Lett. **69**, 1692 (1996)
44. O. Matsuda, O.B. Wright, J. Opt. Soc. Am. B **19**, 3028 (2002)
45. C. Rossignol, B. Perrin, S. Laborde, L. Vandenbulcke, M.I.D. Barros, P. Djemia, J. Appl. Phys. **95**, 4157 (2004)
46. T. Lee, K. Ohmori, C.S. Shin, D.G. Cahill, I. Petrov, J.E. Greene, Phys. Rev. B **71**, 144106 (2005)
47. C. Mechri, P. Ruello, J.M. Breteau, M.R. Baklanov, P. Verdonck, V. Gusev, Appl. Phys. Lett. **95**, 091907 (2009)
48. P.M. Walker, J.S. Sharp, A.V. Akimov, A.J. Kent, Appl. Phys. Lett. **97**, 073106 (2010)
49. L.J. Shelton, F. Yang, W.K. Ford, H.J. Maris, Phys. Stat. Sol. B **242**, 1379 (2005)
50. G. Tas, H.J. Maris, Phys. Rev. B **55**, 1852 (1997)
51. M.E. Msall, O.B. Wright, O. Matsuda, J. Phys. Conf. Ser. **92**, 012026 (2007)
52. T. Pezeril, C. Klieber, S. Andrieu, K.A. Nelson, Phys. Rev. Lett. **102**, 107402 (2009)
53. D. Lim, R.D. Averitt, J. Demsar, A.J. Taylor, N. Hur, S.W. Cheong, Appl. Phys. Lett. **83**, 4800 (2003)

54. C. Rossignol, B. Perrin, B. Bonello, P. Djemia, P. Moch, H. Hurdequint, Phys. Rev. B **70**, 094102 (2004)
55. Y.H. Ren, M. Trigo, R. Merlin, V. Adyam, Q. Li, Appl. Phys. Lett. **90**, 251918 (2007)
56. D. Wang, A. Cross, G. Guarino, S. Wu, R. Sobolewski, A. Mycielski, Appl. Phys. Lett. **90**, 211905 (2007)
57. A.V. Scherbakov, A.S. Salasyuk, A.V. Akimov, X. Liu, M. Bombeck, C. Brüggemann, D.R. Yakovlev, V.F. Sapega, J.K. Furdyna, M. Bayer, Phys. Rev. Lett. **105**, 117204 (2010)
58. F. Decremps, L. Belliard, B. Perrin, M. Gauthier, Phys. Rev. Lett. **100**, 035502 (2008)
59. A. Yamamoto, T. Mishina, Y. Masumoto, M. Nakayama, Phys. Rev. Lett. **73**, 740 (1994)
60. J.J. Baumberg, D.A. Williams, K. Köhler, Phys. Rev. Lett. **78**, 3358 (1997)
61. K. Mizoguchi, M. Hase, S. Nakashima, M. Nakayama, Phys. Rev. B **60**, 8262 (1999)
62. Ü. Özgür, C.W. Lee, H.O. Everitt, Phys. Rev. Lett. **86**, 5604 (2001)
63. K.H. Lin, G.W. Chern, Y.K. Huang, C.K. Sun, Phys. Rev. B **70**, 073307 (2004)
64. N.W. Pu, Phys. Rev. B **72**, 115428 (2005)
65. K.H. Lin, G.W. Chern, C.T. Yu, T.M. Liu, C.C. Pan, G.T. Chen, J.I. Chyi, S.W. Huang, P.C. Li, C.K. Sun, IEEE Trans. Ultrason. Ferroelectr. Freq. Control **52**, 1404 (2005)
66. C.E. Martinez, N.M. Stanton, P.M. Walker, A.J. Kent, S.V. Novikov, C.T. Foxon, Appl. Phys. Lett. **86**, 221915 (2005)
67. R. Liu, G.D. Sanders, C.J. Stanton, C.S. Kim, J.S. Yahng, Y.D. Jho, K.J. Yee, E. Oh, D.S. Kim, Phys. Rev. B **72**, 195335 (2005)
68. O. Matsuda, T. Tachizaki, T. Fukui, J.J. Baumberg, O.B. Wright, Phys. Rev. B **71**, 115330 (2005)
69. R.N. Kini, A.J. Kent, N.M. Stanton, M. Henini, Appl. Phys. Lett. **88**, 134112 (2006)
70. N.D. Lanzillotti-Kimura, A. Fainstein, A. Huynh, B. Perrin, B. Jusserand, A. Miard, A. Lemaître, Phys. Rev. Lett. **99**, 217405 (2007)
71. W.I. Kuo, E.Y. Pan, N.W. Pu, J. Appl. Phys. **103**, 093533 (2008)
72. E.J. Reed, M.R. Armstrong, K.Y. Kim, J.H. Glownia, Phys. Rev. Lett. **101**, 014302 (2008)
73. Y. Wang, C. Liebig, X. Xu, R. Venkatasubramanian, Appl. Phys. Lett. **97**, 083103 (2010)
74. B. Bonello, A. Ajinou, V. Richard, P. Djemia, S.M. Chérif, J. Acoust. Soc. Am. **110**, 1943 (2001)
75. G.A. Antonelli, H.J. Maris, S.G. Malhotra, J.M.E. Harper, J. Appl. Phys. **91**, 3261 (2002)
76. D.M. Profunser, J. Vollmann, J. Dual, Ultrasonics **40**, 747 (2002)
77. M.A. van Dijk, M. Lippitz, M. Orrit, Phys. Rev. Lett. **95**, 267406 (2005)
78. A. Devos, F. Poinsotte, J. Groenen, O. Dehaese, N. Bertru, A. Ponchet, Phys. Rev. Lett. **98**, 207402 (2007)
79. S. Ayrinhac, A. Devos, A.L. Louarn, P.A. Mante, P. Emery, Opt. Lett. **35**, 3510 (2010)
80. A. Bruchhausen, R. Gebs, F. Hudert, D. Issenmann, G. Klatt, A. Bartels, O. Schecker, R. Waitz, A. Erbe, E. Scheer, J.R. Huntzinger, A. Mlayah, T. Dekorsy, Phys. Rev. Lett. **106**, 077401 (2011)
81. G. Tas, J.J. Loomis, H.J. Maris, A.A. Bailes III, L.E. Seiberling, Appl. Phys. Lett. **72**, 2235 (1998)
82. C. Rossignol, N. Chigarev, M. Ducousso, B. Audoin, G. Forget, F. Guillemot, M.C. Durrieu, Appl. Phys. Lett. **93**, 123901 (2008)
83. B. Audoin, C. Rossignol, N. Chigarev, M. Ducousso, G. Forget, F. Guillemot, M.C. Durrieu, Ultrasonics **50**, 202 (2010)
84. H.Y. Hao, H.J. Maris, Phys. Rev. B **64**, 064302 (2001)
85. O.L. Muskens, J.I. Dijkhuis, Phys. Rev. Lett. **89**, 285504 (2002)
86. E. Péronne, B. Perrin, Ultrasonics **44**, e1203 (2006)
87. A.V. Akimov, A.V. Scherbakov, D.R. Yakovlev, C.T. Foxon, M. Bayer, Phys. Rev. Lett. **97**, 037401 (2006)
88. P. Hess, A.M. Lomonosov, Ultrasonics **50**, 167 (2010)
89. O.B. Wright, K. Kawashima, Phys. Rev. Lett. **69**, 1668 (1992)
90. T. Tachizaki, T. Muroya, O. Matsuda, Y. Sugawara, D.H. Hurley, O.B. Wright, Rev. Sci. Instrum. **77**, 043713 (2006)

91. J.D. Choi, T. Feurer, M. Yamaguchi, B. Paxton, K.A. Nelson, Appl. Phys. Lett. **87**, 081907 (2005)
92. A.A. Maznev, K.A. Nelson, J.A. Rogers, Opt. Lett. **23**, 1319 (1998)
93. J. Wang, C. Guo, Phys. Rev. B **75**, 184304 (2007)
94. S. Yamaguchi, T. Tahara, J. Raman Spectrosc. **39**, 1703 (2008)
95. R.J. Smith, M.G. Somekh, S.D. Sharples, M.C. Pitter, I. Harrison, C. Rossignol, Meas. Sci. Technol. **19**, 055301 (2008)
96. S.I. Anisimov, B.L. Kapeliovich, T.L. Perel'man, Sov. Phys. JETP **39**, 375 (1974)
97. O.L. Anderson, in *Physical Acoustics*, vol. 3B, ed. by W.P. Mason (Academic Press, New York, 1965), chap. 2, pp. 43–95
98. A.A. Kolomenskii, A.A. Maznev, Phys. Rev. B **48**, 14502 (1993)
99. R.E. Vines, S. Tamura, J.P. Wolfe, Phys. Rev. Lett. **74**, 2729 (1995)
100. K. Nakano, K. Hane, S. Okuma, T. Eguchi, Opt. Rev. **4**, 265 (1997)
101. M. Clark, S.D. Sharples, M.G. Somekh, J. Acoust. Soc. Am. **107**, 3179 (2000)
102. J.V. Knuuttila, P.T. Tikka, M.M. Salomaa, Opt. Lett. **25**, 613 (2000)
103. J.E. Graebner, B.P. Barber, P.L. Gammel, D.S. Greywall, Appl. Phys. Lett. **78**, 159 (2001)
104. T. Hesjedal, G. Behme, Appl. Phys. Lett. **78**, 1948 (2001)
105. Y. Sugawara, O.B. Wright, O. Matsuda, M. Takigahira, Y. Tanaka, S. Tamura, V.E. Gusev, Phys. Rev. Lett. **88**, 185504 (2002)
106. J.L. Blackshire, S. Sathish, B.D. Duncan, M. Millard, Opt. Lett. **27**, 1025 (2002)
107. Y. Sugawara, O.B. Wright, O. Matsuda, Appl. Phys. Lett. **83**, 1340 (2003)
108. C. Glorieux, K. Van de Rostyne, J.D. Beers, W. Gao, S. Petillion, N.V. Riet, K.A. Nelson, J.F. Allard, V.E. Gusev, W. Lauriks, J. Thoen, Rev. Sci. Instrum. **74**, 465 (2003)
109. D. Shilo, E. Zolotoyabko, Phys. Rev. Lett. **91**, 115506 (2003)
110. J.A. Scales, A.E. Malcolm, Phys. Rev. E **67**, 046618 (2003)
111. A. Miyamoto, S. Matsuda, S. Wakana, A. Ito, Elec. Comm. Jpn. Pt. 2 **87**, 1295 (2004)
112. O.B. Wright, O. Matsuda, Y. Sugawara, Jpn. J. Appl. Phys. **44**, 4292 (2005)
113. D.H. Hurley, K.L. Telschow, Phys. Rev. B **71**, 241410(R) (2005)
114. D.H. Hurley, O.B. Wright, O. Matsuda, T. Suzuki, S. Tamura, Y. Sugawara, Phys. Rev. B **73**, 125403 (2006)
115. D.M. Profunser, O.B. Wright, O. Matsuda, Phys. Rev. Lett. **97**, 055502 (2006)
116. T. Fujikura, O. Matsuda, D.M. Profunser, O.B. Wright, J. Masson, S. Ballandras, Appl. Phys. Lett. **93**, 261101 (2008)
117. D.M. Profunser, E. Muramoto, O. Matsuda, O.B. Wright, U. Lang, Phys. Rev. B **80**, 014301 (2009)
118. N. Wu, K. Hashimoto, K. Kashiwa, T. Omori, M. Yamaguchi, Jpn. J. Appl. Phys. **48**, 07GG01 (2009)
119. B. Bonello, L. Belliard, J. Pierre, J.O. Vasseur, B. Perrin, O. Boyko, Phys. Rev. B **82**, 104109 (2010)
120. T. Saito, O. Matsuda, M. Tomoda, O.B. Wright, J. Opt. Soc. Am. B **27**, 2632 (2010)
121. B.R. Tittmann, L.A. Ahlberg, J.M. Richardson, R.B. Thompson, IEEE Trans. Ultrason. Ferroelectr. Freq. Control **34**, 500 (1987)
122. A.A. Maznev, K.A. Nelson, T. Yagi, Thin Solid Films **290–291**, 294 (1996)
123. C. Glorieux, W. Gao, S.E. Kruger, K. Van de Rostyne, W. Lauriks, J. Thoen, J. Appl. Phys. **88**, 4394 (2000)
124. O. Matsuda, C. Glorieux, J. Acoust. Soc. Am. **121**, 3437 (2007)
125. J. Goossens, P. Leclaire, X. Xu, C. Glorieux, L. Martinez, A. Sola, C. Siligardi, V. Cannillo, T. Van der Donck, J.P. Celis, J. Appl. Phys. **102**, 053508 (2007)
126. B. Taylor, H.J. Maris, C. Elbaum, Phys. Rev. Lett. **23**, 416 (1969)
127. B. Taylor, H.J. Maris, C. Elbaum, Phys. Rev. B **3**, 1462 (1971)
128. G.A. Northrop, J.P. Wolfe, Phys. Rev. Lett. **43**, 1424 (1979)
129. A.G. Every, W. Sachse, Phys. Rev. B **44**, 6689 (1991)
130. Y. Tanaka, M. Takigahira, S. Tamura, Phys. Rev. B **66**, 075409 (2002)
131. M.M. Sigalas, E.N. Economou, J. Sound Vib. **158**, 377 (1992)

132. M.S. Kushwaha, P. Halevi, L. Dobrzynski, B. Djafari-Rouhani, Phys. Rev. Lett. **71**, 2022 (1993)
133. Y. Tanaka, S. Tamura, Phys. Rev. B **60**, 13294 (1999)
134. F. Meseguer, M. Holgado, D. Caballero, N. Benaches, J. Sánchez-Dehesa, C. López, J. Llinares, Phys. Rev. B **59**, 12169 (1999)
135. S. Benchabane, A. Khelif, J.Y. Rauch, L. Robert, V. Laude, Phys. Rev. E **73**, 065601(R) (2006)

Chapter 8
Sate-of-the-Art of Terahertz Science and Technology

Masayoshi Tonouchi

8.1 Introduction

Terahertz (THz) science and technology is a new research field. THz refers to the frequency gap between the infrared and microwaves, typically the frequency range from 100 GHz to 30 THz. In fact, studies in the THz range have long been conducted in fields such as astronomy and analytical science. However, recent innovations in THz technologies are leading to a wide variety of applications in the field of Information and communications technology (ICT); biology and medical sciences, THz nondestructive evaluation (THz-NDE); homeland security, quality control of food and agricultural products; global environmental monitoring, and ultrafast computing, among others [1–3].

Why is the THz science and technology so important? The reasons are that it provides a novel sensing technique for spectroscopy and imaging in the THz frequency range; secondary, innovative fundamentals are for massive information and communication technology in future generations; as third, one can expect new science; in addition, we need standard measure for application. As for the sensing, THz waves are characterized by their low photons energy close to biofluctuation level. Besides, the THz waves are harmless in comparison with X-ray beam. The THz waves have relatively high transmittance of the matters compared with optical beam, and have smaller resolution for the imaging compared with micrometer waves, which enable us to distinguish specific materials buried in different. One of the important research trends in the field of electronics is to develop ultra-high-speed devices operative at THz frequency range, following to logic circuits at a clock rate over 100 GHz and wireless communication technology with 10 Gbps bit rate being achieved. For electronic devices in next generation, one has also to know basic/dielectric properties of the materials at a THz frequency range for high speed device development. Another

M. Tonouchi (✉)
Institute of Laser Engineering, Osaka University, 2-6 Yamada-Oka, Suita, Osaka 565-0871, Japan
e-mail: tonouchi@ile.osaka-u.ac.jp

K. Shudo et al. (eds.), *Frontiers in Optical Methods*,
Springer Series in Optical Sciences 180, DOI: 10.1007/978-3-642-40594-5_8,
© Springer-Verlag Berlin Heidelberg 2014

interest is to study ultrafast phenomena in physical matters. Femtosecond (fs) lasers have opened new research field to pursue transient response in those. Especially THz waves probes low energy dynamics, which gives essential information of many materials such as electron scattering in semiconductors, molecular vibrations in bio-materials, and so on. Such transient low energy dynamics is unexplored area of research. Realistic application requires standards of THz waves and regulation for Electro Magnetic Compatibility (EMC). Above frequencies around 300 GHz, the waves are not allocated for commercial use. We also need to discuss environmental and astronomical use of THz waves. From a viewpoint of above interests, scientific research of THz waves is absolutely imperative for future applications.

Although one always has interested in the unexplored spectrum region, recent breakthroughs have given the research a boost up to new level. THz time-domain spectroscopy (THz-TDS), developed in the mid 1980s, represented a major break-through and was followed by the invention of THz imaging based on THz-TDS. THz-TDS provides an easy measure to evaluate complex refractive index of the materials [4]. Various types of THz research and applications have been simulated using THz imaging [5]. Other breakthroughs are THz quantum cascade lasers (THz-QCL) [6], bright THz source [6], and THz electronic devices, and so on. THz-QCL is an intense THz electronic source, and bright sources such as parametric/Cherenkov THz bright source can even induce nonlinear response of the materials with strong THz field. A number of electronic THz devices such as resonant tunneling diode (RTD), uni-travel-carrier photodiodes, and a single flux quantum circuit are under development.

Important platforms involving THz are classified into THz sources, THz detec-tors, modulators/manipulators, imaging methods, basic science such as interaction between THz waves and materials, application trials, THz standards/ EMC, develop-ment of databases, and so on. This article reviews the recent progress in THz science and technology and provides an introduction to some of the recent topics in THz applications.

8.2 New Twist in THz Science and Technology

8.2.1 THz Sources

Owing to the breakthroughs in the field of THz sources and time-domain spec-troscopy, THz research has been triggered among many fundamental studies. Figure 8.1 shows the THz emission power as a function of frequency. There are three major approaches for developing THz sources. The first approach is optical THz generation; the second, the recently developed THz quantum cascade laser (QCL); and the third, the uses of solid-state electronic devices.

THz waves can be generated by optical methods, e.g., by an ultrafast photocurrent generated in a photoconductive switch or semiconductor by using electric-field

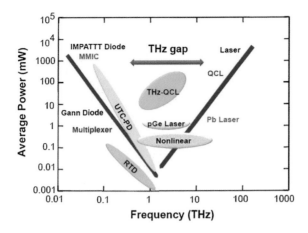

Fig. 8.1 THz emission power from various devices, plotted as a function of frequency

carrier acceleration or the photo-Dember effect, and by nonlinear optical effects such as optical rectification, difference-frequency generation, or optical parametric oscillation (OPO). A steady progress has been made in the former method, e.g., the development of photoconductive switches adapted for excitation at a wavelength of 1.5 μm [7, 8]. Photomixing is also a potential technique for THz beam generation. Among the candidates suitable for photomixers, uni-travel-carrier photodiodes (UTC-PDs) are promising devices. Itoh et al. have succeeded generating THz power exceeding 20 μW at 1 THz, which is suitable for short-range wireless communication [9].

Very recent topics in the field of nonlinear optical-THz conversion are THz Cherenkov radiation [10, 11] and air plasma generation [12]. The emission mechanism for the former is explained as follows. When fs optical pulses travel in a LiNbO$_3$ crystal, THz waves are generated by parametric fluorescence at an angle of approximately 64 degrees with the direction of propagation of the incident pulse. However, THz waves are strongly absorbed in LiNbO$_3$. Therefore, in order to avoid such absorption, Hebling et al. introduced fs optical pulses with a tilted wave front so as to excite THz waves at the wavefront of the Cherenkov radiation as shown in Fig. 8.2; use of a tilted pulse front enhances THz wave generation. Nagai et al. have improved the system and realized high-field generation of monocycle THz pulses above 1 MV/cm. Figure 8.3 shows an example for the emission waveform and corresponding spectrum [13, 14].

Air/gas plasma also can produce a high-power THz beam. In addition to the high power, it covers wide broad bands up to 30 THz. The emissivity of plasma strongly depends on gaseous species, and increases with a decrease in the ionization potential. Thus, Xe gas can produce THz beams having much higher power than those produced by He gas.

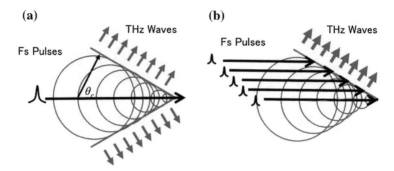

Fig. 8.2 **a** Cherenkov radiation in LiNbO₃ crystal and **b** improved THz generation using fs pulses with a tilted wave front

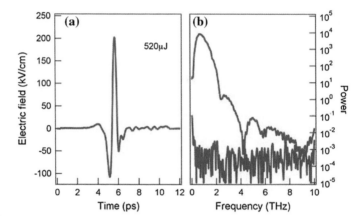

Fig. 8.3 **a** Temporal THz waveform and **b** corresponding Fourier spectrum

Remarkable progress has been achieved in the development of THz-QCL. A state-of-the-art THz-QCL operates at a lowest frequency of 0.68 THz and a temperature of 225 K with an external magnetic field, and it provides a power of 140 and 250 mW under continuous-wave (CW) and pulsed operations, respectively [15, 16].

Electronic generation of THz waves is one of the key technologies for ICT to realize mobile, robust, and inexpensive systems. Very recently, Asada et al. succeeded in generating THz waves having frequencies above 1 THz from resonant tunneling diodes (RTDs). Figure 8.4 shows an example of RTD and emission spectrum [17]. Further, Otsuji et al. have developed plasmon-resonant THz emitters and succeeded in producing broadband THz waves up to 7 THz [18, 19]. These solid-state devices can be used in a variety of applications.

(a)   **(b)**

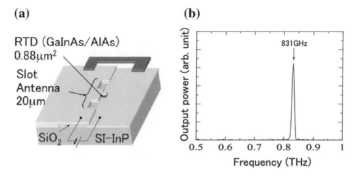

Fig. 8.4 **a** Schematic of RTD. **b** THz emission at 832 GHz from RTD

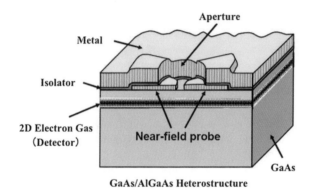

Fig. 8.5 Schematic of all-in-one-chip near-field THz detector

8.2.2 THz Detectors

Continuous efforts have been made to develop high-quality THz detectors. With regard to recent advances, Kawano et al., have developed a near-field THz detector, which consists of a two-dimensional (2D) electron gas THz detector, a small aperture, and a near-field probe antenna fabricated in a single chip, as shown in Fig. 8.5 [20]. This detector enables us to obtain a high-resolution image of a THz field. They are also developing a novel THz detector that utilizes graphene operating under a magnetic field [21].

Recently, Oda et al. have optimized the operational frequency of infrared focal plane imaging arrays [22] operating at a frequency of approximately 3 THz and reduced their NEP to as small as 40 pW at a flame rate of 60 Hz for an image resolution of 320×240 pixels (Fig. 8.6). They optimized the resistance of the absorption layer and window materials. At the moment, this demonstrates the highest performance of a THz camera [23–25].

In addition to imaging arrays, THz applications require many other components such as THz fibers [26], waveguides, and metamaterials [27]. Recently, porous fibers

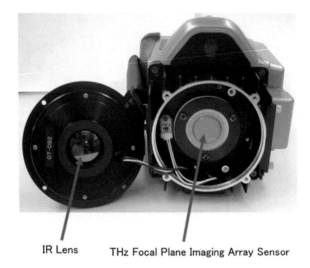

IR Lens   THz Focal Plane Imaging Array Sensor

Fig. 8.6 Developed THz camera

are found to be suitable low-dispersion media for fabricating THz waveguides. In addition, carbon nano-tubes are found to be excellent THz wave polarizers [28].

8.2.3 THz Spectrometer and Imaging Systems

Continuous efforts have been made to develop compact, high-sensitive, and robust system for THz spectroscopy and imaging. As a source of TDS, organic crystals have high optical-to-terahertz conversion efficiency. DAST is one of the promising materials. Otsuka Electronics Co. Ltd. has recently commercialized a compact THz-TDS system that employs an fs fiber laser operating at a wavelength of 1,560 nm, and DAST, which covers a frequency range of up to 7 THz (Fig. 8.7). Since a 1.5 μm

Fig. 8.7 Developed
THz-TDS system

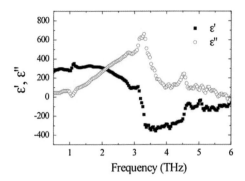

Fig. 8.8 **a** Real and **b** imaginary parts of dielectric constant of an STO film formed on MgO substrate

laser satisfies the phase-matching conditio at around 2.2 THz [29], the conversion efficiency becomes maximum at frequencies above 2 THz, which enables us to utilize THz waves at high frequencies. Figure 8.8 shows an example of the evaluation of the dielectric parameters of $SrTiO_3$ thin films. These 370-nm-thick films grown by laser ablation exhibit a resonance at 3.3 THz, which is attributed to LO photon scattering [30, 31]. These results indicate that one can evaluate the optical constant of a film up to a frequency of 6 THz.

We are studying a unique application of THz waves. Femtosecond laser illumination induces THz wave generation from various types of materials [32]. The observation of THz waveforms enables us to explore the ultrafast nature of electronic materials and devices by employing THz emission spectroscopy. Thus, a laser-THz emission microscope (LTEM) would be a new tool for material science and application [33]. An LTEM system is illustrated schematically in Fig. 8.9. At present, we have obtained the LTEM resolution down to 0.6 μm. As explained later, LTEMs have many applications such as imaging of supercurrent distributions [34], ferroelectric domain [35], and so on.

8.3 New Trends in THz Applications

8.3.1 Conservation of Historic and Artistic Works

There have been innumerable applications of THz-TDS [1]. Fukunaga et al. have recently introduced THz-NDE for conservation of historic and artistic works, which gives us strong impact [23–25, 36]. She observed that different color pallets have different responses to THz waves. This response is much more sensitive than that with light from the mid- and near-infrared regions. Thus, THz-NDE is a potential tool to analyze artwork in order to obtain information that is useful for restoration and conservation.

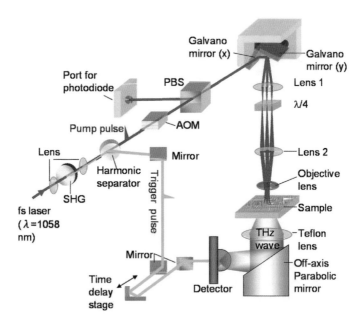

Fig. 8.9 Schematic of LTEM

Figure 8.10 shows one of the observation parts in the Polyptych by the Badia (ca 1,300, Giotto, Uffizi Gallery). The existence of a gold foil under pigments was clearly observed at the outline of the head and wings of the Angel. The layered structure of the painting was obtained non-invasively and without contact. Further, two gesso layers proved that this work was made by using a traditional medieval technique. The information obtained here by using THz imaging is practically useful

Fig. 8.10 Visible (**a**) and THz (**b**) images of a part of Polyptych by the Badia, and its tomographic view (**c**)

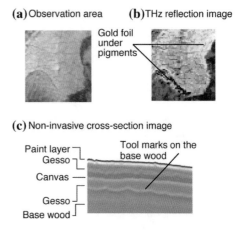

for conservators and would never be obtained using conventional methods. Note that
these techniques are also available for many other NDEs.

8.3.2 THz Emission Spectroscopy

One can observe THz emission from various materials upon femtosecond laser illumination, which reflects a dynamic response of optically excited materials. Thus, THz
emission spectroscopy is one of the potential tools to study material physics. We have
discovered a unique THz emission from $BiFeO_3$ (BFO) [37]. The emission mechanism is attributed to the direct optical modulation of spontaneous polarization P_s.
BFO is well known to poses strong P_s. Generally, THz emission is attributed to real
space carrier acceleration/deceleration, or nonlinear down conversion due to optical rectification/difference frequency generation DFG/parametric generation through
the $\chi^{(2)}$ or $\chi^{(3)}$ effect. Unlike conventional mechanisms, THz generation from BFO
is explained by the optical annihilation of P_s [38].

Figure 8.11a shows an example of THz emission from a photoconductive antenna
made of BFO and Au electrodes. We use an fs laser at a wavelength of 400 nm doubled
by SHG. BFO is poled once by an external electric field and the emission is observed
after removing the field. The dependence of the maximum amplitude on the poling
field shows a clear hysteresis, representing Ps of the BFO film (Fig. 8.11b). These
results indicate that poling performed using illumination from fs laser pulses assists

Fig. 8.11 a Example of THz
emission waveform radiated
from BFO photoconductive
switch, and **b** dependence of
maximum THz amplitude on
poling field. *Solid squares*
represent the data poled with
illumination; *open squares*
and *solid circles*, represent
those poled without illumination

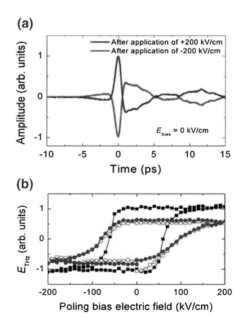

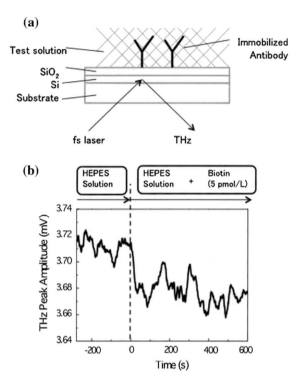

Fig. 8.12 **a** Schematic of sensing chip and **b** time-dependent THz emission amplitude

spontaneous polarization switching, whereas no saturation of poling is observed below an external field of less than 200 kV/cm. This proves that THz emission spectroscopy can be an effective tool for the noncontact and nondestructive evaluation of P_s.

Kiwa et al. have developed a new application of THz emission to study chemical reactions and bio applications [39, 40]. Figure 8.12a shows a schematic of the sensing chip developed by Kiwa et al. SiO_2 and Si thin films were prepared on a substrate. The antibody was immobilized on the SiO_2 surface. When a femtosecond laser pulse hits the Si film from the backside of the chip, Hz waves are generated and radiated toward the free space as a result of the photo-Dember effect. If the antibody is combined with an antigen, the surface potential of the chip shifts and the peak amplitude of THz changes. Thus, the reaction of the antibody and the antigen can be measured by monitoring the peak amplitude of THz. Figure 8.12b shows the peak amplitude of the THz emitted from the chip. Avidin, which is strongly combined with biotin, is immobilized on the chip. A solution of 4-(2-hydroxyethyl)-1-piperazineethanesulfonic acid (HEPES) is applied as the test solution. The THz amplitude rapidly decreases on injecting 5-pmol/L-biotin in the solution. This result indicates that high-sensitive label-free detection of the reaction of proteins can be carried out using the sensing chip. This is the first demonstration of the detection of proteins combinations in water solutions performed using THz technology.

8.3.3 LTEM Application to LSI Defect Analysis

We have developed an LTEM that employs excitation laser pulses at a wavelength of 1.06 μm for the inspection and localization of electrical failures in large-scale integrated (LSI) circuits with multilayered interconnection structures. The LTEM system enables us to record THz emission images from the backside of an LSI chip with a multilayered interconnection structure that prevents observation from the front side. By comparing the recorded THz emission images, we can successfully distinguish a normal circuit from damaged ones with different positions of the interconnection defects, without having to perform any electrical probing [41].

Schematic illustrations of test circuits composed of Si pMOSFETs are shown in Fig. 8.13a, these circuits are evaluated from the backside of the chip. In this figure, the normal circuit is labeled as MOS-N, and the three defective ones are labeled as (MOS-D1-MOS-D3); the defective circuits are prepared with a different position of the disconnection of the line between the heavily doped n-type areas and the electrode pad, as shown by circles in the figure. MOS-D1 and MOS-D2 have disconnections on

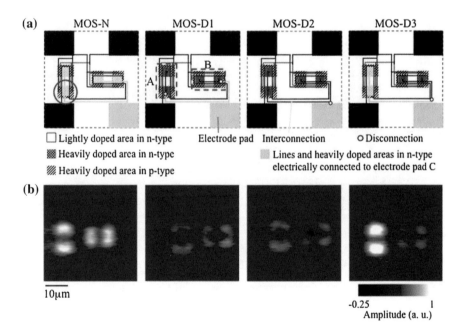

Fig. 8.13 **a** Illustrations of the normal circuit (MOS-N) and the defective ones (MOS-D1-MOS-D3) with different positions of disconnection. **b** Corresponding THz emission images

the lines from both areas A and B, differing in the fact that the lines in MOS-D1 are interrupted near A and B, while in MOSD2, there is a single interruption near the pad C. On the other hand, MOS-D3 has a disconnection only on the line from the area B. The THz emission images of these test circuits were recorded from the backside of the chip and are shown in Fig. 8.13b. By comparing the THz emission images in Fig. 8.13b, one can distinguish the normal circuit from the defective ones. It also can be seen that the THz emission signals from p-n junctions change depending on the position of the line disconnection. MOS-D1 and MOS-D2 radiate the smaller THz emission signals from heavily doped n-type areas in both areas A and B than those emitted by MOS-N; this is due to the disconnections in MOS-D1 and MOS-D2. In the case of MOS-D3, the THz emission signals originating from the heavily doped n-type areas A increase, however, those originating from B decrease because of the disconnection between the electrode pad C and the area B. This result suggests that the THz emission signal increases due to an increase in the transient photocurrent flowing from A into the electrode pad enhances. Further, it can also be seen that there is little influence of the disconnections on the THz emission signals from heavily doped p-type areas. These results suggest that the interconnections and the electrode pad enhance the THz emission efficiency by functioning as an antenna and that the changes in THz emission images are useful for the localization of p-n junctions with Interconnection defects in circuits.

8.4 Summary

THz technology is a cross-cutting one which provides an opportunity to accrete various fields of science and technology, as nano-technology is. Thus, the implementation of the THz research would bring out vast amount of new applications. In this review, recent progress of terahertz fundamental technology and challenge for new application have been introduced. THz sources have been improved drastically by TDS, Cherenkov radiation, air plasma, and QCL's. New types of detectors and camera are also developed. LTEM is introduced as a potential application tool for LIS failure analysis, and others. Art conservation by THz-TDS also gathers much attention by the experts. Many other NDE and imaging applications are expected to be developed soon. THz wireless communication and bio-applications are expected to be used in a large number of commercial applications; however, a long-term strategy is needed to realize such applications.

Acknowledgments Auhtors are grateful to Profs. Asada of Tokyo Institute of Technology, Tanaka of Kyoto University, Nagai of Osaka University, Kiwa of Okayama University, and Drs. Oda of NEC, Fukunaga of NICT, Yamashita of RIKEN, and Kawano of RIKEN for their providing materials.

References

1. M. Tonouchi, Cut. Edge Terahertz Technol. Nat. Photonics **1**, 97–105 (2007)
2. M. Tonouchi, *Terahertz Technology* (Ohmsha, Tokyo, 2006) (in Japaneşe)
3. M. Tonouchi, OYOBUTSURI, textbf75, 60 (2006) (in Japanese)
4. B. Ferguson, X.-C. Zhang, Materials for terahertz science and technology. Nat. Mater. **1**, 26 (2002)
5. D. Mittleman (ed.), *Sensing with Terhaertz Radiation* (Springer, Berlin, 2003)
6. R. Kohler et al., Terahertz semiconductor-heterostructure laser. Nature **417**, 156–159 (2002)
7. M. Suzuki, M. Tonouchi, Fe-implanted InGaAs terahertz emitters for 1.56 μm wavelength excitation. Appl. Phys. Lett. **86**, 051104 (2005)
8. M. Suzuki, M. Tonouchi et al., Excitation wavelength dependence of terahertz emission fronm semiconductor surface Appl. Phys. Lett. **89**, 091111 (2006)
9. T. Nagatsuma, H. Ito, T. Ishibashi, High-power FR photodiode and their applications. Laser Photon. Rev. **3**, 123–137 (2009)
10. J. Hebling, A.G. Stepanov, G. Almasi, B. Bartal, J. Kuhl, Tunable THz pulse generation by optical rectification of ultrashort laser pulses with tilted pulse fronts. Appl. Phys. B **78**, 593–599 (2004)
11. J. Hebling et al., Generation of high-power terahertz pulses by tilted-pulse-front excitation and their application possibilities. J. Opt. Soc. Am. B **25**, B6–B19 (2008)
12. Y. Chen, M. Yamaguchi, M. Wang, X.-C. Zhang, Terahertz pulse generation from noble gases App. Phys. Lett. **91**, 251116 (2007)
13. M. Nagai et al., Broadband and high power terahertz pulse generation beyond excitation bandwidth limitation via $\chi^{(2)}$ cascaded processes in LiNbO$_3$. Opt. Express **17**, 11543 (2009)
14. M. Jewariya, M. Nagai, K. Tanaka, Enhancement of terahertz wave generation by cascaded $\chi^{(2)}$ processes in LiNbO$_3$. J. Opt. Soc. Am. B **26**, A101 (2009)
15. A.W.M. Lee et al., High-power and high-temperature THz quantum-cascade lasers based on lens-coupled metal-metal waveguides. Opt. Lett. **32**, 2840–2842 (2007)
16. A. Wade et al., Magnetic-field-assisted terahertz quantum cascade laser operating up to 225 K. Nat. Photon. **3**, 41–45 (2009)
17. S. Suzuki, et al.:Room-temperature fundamental oscillation of RTD at 831GHz. Appl. Phys. Express **2**, 054501 (2009)
18. Y.M. Meziani et al., Room temperature terahertz emission from grating coupled two-dimensional plasmons. Appl. Phys. Lett. **92**, 201108 (2008)
19. T. Nishimura, N. Magome, H. Kang, T. Otsuji, Spectral Narrowing Effect of a Novel Super-Grating Dual-Gate Structure for Plasmon-Resonant Terahertz Emitter. IEICE Trans. Electron. **E92C**, 696–701 (2009)
20. Y. Kawano, K. Ishibashi, An on-chip near-field terahertz probe and detector. Nat. Photon. **2**, 618–621 (2008)
21. Y. Kawano, Wide-band frequency-tunable terahertz and infrared detection with graphene. Nanotechnol. **24**, 214004 (2013)
22. S. Tohyama et al., New thermally isolated pixel structure for high-resolution (640 X 480) uncooled infrared focal plane arrays. Opt. Eng. **45**, 014001 (2006)
23. K. Fukunaga et al., Real-time terahertz imaging for art conservation science. J. Euro. Opt. Soc. **3**, 08027 (2008)
24. K. Fukunaga, I. Hosako, I.N. Durling, M. Picollo, Terahertz imaging systems: a non-invasive technique for the analysis of paintings. Proc. SPIE **7391**(73910D) (2009)
25. K. Fukunaga, Non-destructive THz imaging of a Giotto masterpiece IIC News in Conservation, February issue, p. 2 (2009)
26. S. Atakaramians et al., THz porous fibers: design, fabricatio and experimental characterization. Opt. Express **17**, 14053–14062 (2009)
27. H.T. Chen et al., Active Terahertz Metamater. Dev. Nat. **444**, 597–600 (2006)
28. L. Ren et al., Carb. Nanotub. Terahertz Polarizer Nano Lett. **9**, 2610 (2009)

29. M. Yoshimura et al., Growth of 4-dimethylamino-N-methyl-4-stilbazolium tosylate (DAST) crystal and its application to THz wave generation 17PS-24, in *Ext. Abs. International Workshop Terahertz Technology, Osaka, 2005*

30. M. Misra et. al., Observation of TO1 Soft Mode in $SrTiO_3$ Film by Terahertz Time Domain Spectroscopy. Appl. Phys. Lett. **87**, 182909 (2005)

31. R. Kinjo et. al., Observation of strain effects of SrTiO3 thin films by terahertz time-domain spectroscopy with a 4-dimethylamino-N-methyl-4-stilbazolium tosylate emitter. Jpn. J. Appl. Phys. **48** (2009) (in press)

32. N. Kida, H. Murakami, M. Tonouchi, in *Terahertz optics in strongly correlated electron systems* ed. by K. Sakai. Terahertz Optoelectronics (Springer, Berlin, 2005)

33. S. Kim, H. Murakami, M. Tonouchi, Transmission-type laser THz emission microscope using a solid immersion lens. IEEE J. Select Topic Quant. Electron. **14**, 498 (2008)

34. M. Tonouchi, M. Yamashita, M. Hangyo, Terahertz radiation imaging of supercurrent distribution in vortex-penetrated $YBa_2Cu_3O_{7-\delta}$ thin film strips. J. Appl. Phys. **87**, 7366–7375 (2000)

35. D.S. Rana et al., Visualization of photoassisted polarization switching and its consequences in $BiFeO_3$ thin films probed by terahertz radiation. Appl. Phys. Lett. **91**, 031909 (2007)

36. K. Fukunaga et al., Terahertz spectroscopy for art conservation IEICE Electron. Exp.**4**, 258–263 (2007)

37. K. Takahashi, N. Kida, M. Tonouchi, Terahertz radiation by an ultrafast spontaneous polarization modulation of multiferroic $BiFeO_3$ thin films. Phys. Rev. Lett. **96**, 117402 (2006)

38. D.S. Rana, I. Kawayama, K.R. Mavani, K. Takahashi, H. Murakami, M. Tonouchi, Understanding the nature of ultrafast polarization dynamics of ferroelectric memory in the multiferroic $BiFeO_3$ thin films. Adv. Mater. **21**, 2881–2885 (2009)

39. T. Kiwa et al., Chemical sensing plate with a laser-terahertz monitoring system. Appl. Opt. **47**, 3324 (2008)

40. T. Kiwa et al., A Terahertz chemical microscope to visualize chemical concentration in microfluidic chip. Jpn. J. Appl. Phys. (Exp. Lett.) **46**, L1052 (2007)

41. M. Yamashita et al., Backside observation of large-scale integrated circuits with multilayered interconnections using laser terahertz emission microscope. Appl. Phys. Lett. **94**, 191104 (2009)

Chapter 9
Broadband Terahertz Spectroscopy of Thin Films

Ikufumui Katayama and Masaaki Ashida

In this chapter, recent progress in Terahertz time-domain spectroscopy (THz-TDS) is reviewed, focusing on broadening of the detection bandwidth. By using an ultrashort pulsed laser with the pulse duration of 15 fs and Photoconducting antennas in reflection geometry, it is now possible to construct broadband terahertz spectrometers that are capable of measuring the real and imaginary parts of the Dielectric constants of thin films over a wide frequency range. The detection bandwidth has been expanded to 20 THz with continuous phase information, except for a small discontinuity around 8 THz. Even higher frequencies such as 170 THz can be generated and detected using a 5-fs pulsed laser with organic Nonlinear crystal and photo-conducting antennas. These methods have a wide range of applications and several examples, such as characterization of dielectric thin films, and Light-pump terahertz probe spectroscopy are reviewed.

9.1 Introduction

Development of ultrashort pulsed laser technology enables us to observe photoinduced phenomena and photoreactions with extremely high time resolution, which was impossible in the past. The high electric field of the ultrashort laser pulses is useful for nonlinear spectroscopy, and has contributed to a better understanding of condensed matter physics [1, 2]. Recently, it was demonstrated that the high time

I. Katayama (✉)
Graduate School of Engineering, Yokohama National University, Tokiwadai 79-5,
Hodogay, Yokohama 240-8501, Japan
e-mail: katayama@ynu.ac.jp

M. Ashida
Graduate School of Engineering Science, Osaka University, Machikaneyamacho 1-3,
Toyonaka 560-8531, Japan
e-mail: ashida@mp.es.osaka-u.ac.jp

K. Shudo et al. (eds.), *Frontiers in Optical Methods*,
Springer Series in Optical Sciences 180, DOI: 10.1007/978-3-642-40594-5_9,
© Springer-Verlag Berlin Heidelberg 2014

resolution achieved with lasers can be applied to the detection of high-frequency electric fields, namely terahertz waves. This so-called terahertz technology has attracted much attention [3–6].

The terahertz region lies between the microwave and infrared frequencies (1 THz $= 10^{12}$ Hz $= 300 \, \mu m = 33 \, cm^{-1} = 4.1$ meV $= 48$ K), and is difficult to access by conventional optical and electronic technologies. This is because the energy of the terahertz wave is lower than that of thermal radiation at room temperature. Therefore, terahertz detectors are often strongly affected by thermal fluctuations. However, D. H. Auston et al. demonstrated in 1984 that laser pulses with pulse duration below 1 ps can be used to generate and detect coherent electromagnetic waves in the terahertz region [3], and after their work, C. Fattinger and X. C. Zhang demonstrated that terahertz radiation can be used in free space. These studies triggered extensive terahertz research using ultrashort laser pulses [7, 8].

This new method for generation and detection of terahertz waves uses coherent electromagnetic waves, and therefore, is little affected by thermal fluctuation and can easily be applied to measurements performed at room temperature. The terahertz region lies between the infrared region where ionic vibrations are observed and the microwave region where the relaxational modes are observed, and is important because excitations related to the phase transitions of materials, such as ferroelectric soft-mode, magnetic excitations, and collective electronic excitations, exist in this region [9, 10]. Furthermore, the vibrational motions of large molecules such as proteins normally exist in the terahertz region, and detection of transformational modes is expected to be useful for understanding protein functions [11]. In addition, using the property that many materials, including papers and plastics, are transparent to terahertz waves, application of terahertz technology for the imaging spectroscopy has been pursued. Since the energy of terahertz waves is much lower than that of X-rays, safe, nondestructive and complementary applications to the imaging of living cells and the interior of materials are expected [12–14]. In this chapter, we briefly describe the principles of terahertz spectroscopy using ultrashort laser pulses, and focus on research to extend the bandwidth of terahertz spectroscopy, including our studies. For a broader review of terahertz spectroscopy, excellent books on the terahertz technologies are available [6, 14]. Other chapters in this book also deal with cutting-edge research on terahertz responses of materials and various applications. We will also briefly describe a new spectroscopic technique, light pump-terahertz probe spectroscopy, which has been made possible by using pulsed terahertz sources.

9.2 Generation and Detection of Terahertz Electric Fields

First, we will introduce the principles of generation and detection of terahertz electric fields, particularly from the viewpoint of broadening the bandwidth. There are two major methods to generate and detect terahertz waves: one is the use of a **Photoconducting antenna** that picks up transient currents associated with the photoexcitation of a semiconductor substrate, and the other is **Optical rectification**, which uses the

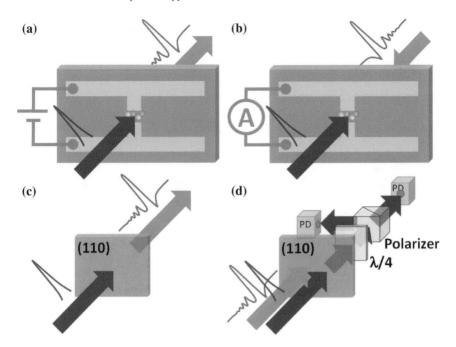

Fig. 9.1 **a** Schematic of terahertz generation using a photo-conducting antenna. High voltage is applied between the two electrodes. **b** Schematic of terahertz detection using a photo-conducting antenna. The current between two electrodes is measured. **c** Schematic of terahertz generation using optical rectification. **d** Schematic of terahertz detection using electro-optic sampling. The polarization rotation is measured with a quarter waveplate, polarizer and two photodiodes

same principle as **Difference frequency generation**. Each method is illustrated schematically in Fig. 9.1.

There are of course many other techniques to generate and detect terahertz electric fields. For example, use of photoinduced plasma offers an interesting approach to broadband and long-distance detection [15–18]. The surface electric field in semi-conductors can also be utilized to generate terahertz waves [8]. These are outside the scope of this chapter; however, they are covered in some of the cited references.

9.2.1 Photoconducting Antenna

The photoconducting antenna was first used in 1984. A gold electrode with the shape of an antenna is deposited on a semiconductor substrate. A DC voltage is applied and an ultrashort laser pulse is focused between these electrodes. The terahertz electric field is generated because of the change in the current induced in the semiconductor [3]. The current that flows between the electrodes can be described by the following equation:

$$j = n(t)e\mu F. \tag{9.1}$$

Here, μ is the mobility of the carriers, e the charge of the photoexcited carriers, $n(t)$ the carrier density, F the applied DC electric field. In semiconductors with high resistivity, $n(t)$ is nearly equal to 0 when the laser light is not applied, and therefore, no current is induced between the electrodes even at high voltage. When the antenna is illuminated with laser pulses whose intensity changes with time as $I(t)$, the time dependence of the carrier density is given by,

$$n(t) \propto \int_{-\infty}^{t} dt' I(t') e^{-(t-t')/\tau}, \tag{9.2}$$

when the lifetime of the cariers is τ. According to the Maxwell equation, the transient current produces electromagnetic waves whose waveform changes according to the current. The waveform of the electric field is proportional to the time derivative of the current as follows:

$$E(t) \propto \frac{dj}{dt}. \tag{9.3}$$

Fourier transformation gives the spectrum of the generated electric field as

$$E(\omega) \propto i\omega j(\omega) \propto i\omega n(\omega) \propto i\omega I(\omega) \frac{\tau}{1 - i\omega\tau}. \tag{9.4}$$

The last term is the Fourier transform of the exponential decay of photoexcited carriers. The spectrum has the shape of a Debye dispersion. In the calculation we used the folding theorem of Fourier transformation. Equation 9.4 indicates that the bandwidth of the generated electromagnetic wave is determined mainly by the carrier lifetime and pulse duration of the laser. For example, the low-temperature-grown GaAs is very efficient in generating electromagnetic waves at 1 THz because it has a very short lifetime of about 1 ps (10^{-12} s) [19].

At the detector, a gating laser pulse is directed to the photoconducting antenna, on which the terahertz wave under investigation is also focused. The current induced by the terahertz electric field is observed as a signal. Here, we write the electric field of the terahertz wave as $E(t')$, and the time delay between the terahertz wave and the laser pulse as t. Then the current observed is given by

$$j(t) = \int_{-\infty}^{\infty} dt' n(t' - t) e\mu E(t'). \tag{9.5}$$

We measure this induced current as a function of the time delay t, and the obtained waveform is Fourier-transformed to get the frequency spectrum. Using the folding theorem, the spectrum can be written as

$$J(\omega) \propto n(\omega) E(\omega) \propto E(\omega) I(\omega) \frac{\tau}{1 - i\omega\tau}. \tag{9.6}$$

Therefore, the obtained spectrum is proportional to the terahertz spectrum with some sensitivity factors. The sensitivity is again higher when the photoexcited carriers have shorter lifetime and the laser pulse has shorter duration. The sensitivity also depends on the design of the antenna, and by tuning the shape, we can adjust the sensitivity, especially at low frequencies, because the size of the antenna is almost the same as the wavelength of the incident terahertz waves [20, 21].

9.2.2 Optical Rectification and Electro-Optic Sampling

Electrooptical crystals are also widely used material to generate and detect coherent electric fields. In this case, a second-order nonlinear process, namely differential frequency generation, is utilized. The second-order nonlinear polarization can be written as follows, using the second-order **Nonlinear susceptibility** of $\chi_2(\omega_1, \omega_2; \omega_2 - \omega_1)$:

$$P_2 = \iint d\omega_1 d\omega_2 \chi_2(\omega_1, \omega_2; \omega_2 - \omega_1) E^*(\omega_1) E(\omega_2) e^{i(\omega_2 - \omega_1)t}. \qquad (9.7)$$

Here, $E(\omega)$ is the complex Fourier spectrum of the incident laser. If the energy of the laser is sufficiently low compared to the absorption edge of the electrooptic crystal, the nonlinear coefficient can be regarded as a constant. Then, second-order polarization is given by

$$P_2 \propto \iint d\omega_1 d\omega_2 E^*(\omega_1) E(\omega_2) e^{i(\omega_2 - \omega_1)t} \propto |E(t)|^2. \qquad (9.8)$$

Therefore, the difference frequency generation corresponds to rectification of the laser pulse, which is called **Optical rectification**. As the pulse duration becomes shorter, the change in the electric field becomes faster, which means that a broadband terahertz wave can be generated. In practice, the terahertz absorption of the electrooptic crystal itself and phase mismatch between the lasers and terahertz wave affect the bandwidth of the generated terahertz wave [22]. These effects can be avoided if we use thin crystals, which are normally used to generate terahertz waves with broad bandwidth.

For detection, the inverse process of the electrooptic effect is used. The nonlinear polarization can be written as

$$P_2 = \iint d\omega_1 d\omega_2 \chi_2(\omega_1, \omega_2; \omega_2 \pm \omega_1) F(\omega_1) E(\omega_2) e^{i(\omega_2 \pm \omega_1)t}. \qquad (9.9)$$

Here, $F(\omega)$ is the terahertz electric field. If it is assumed that the nonlinear coefficient is constant, the nonlinear polarization is proportional to the electric field of the incident laser. Therefore, it can be regarded as representing the change in dielectric constant (or refractive index). Within a linear approximation, the change in refractive index is proportional to the terahertz electric field. When we use an isotropic nonlinear crystal for detection, the terahertz electric field induces anisotropy in the refractive index, which can be observed as a polarization rotation. The polarization rotation is normally measured with a quarter waveplate and polarizer; this is the so-called **Electro-optic sampling** method.

In order to get broad bandwidth, it is required to use thin electrooptic crystals, as in the case of generation. However, sensitivity to the electric field is smaller at lower thickness. Furthermore, multiple reflections in the crystal affect the transmitted waveform, and the analysis becomes difficult as this becomes more significant. In order to avoid multiple reflection, the thick electrooptic crystal insensitive to the electric field is attached to the thin electrooptic crystal that is used to observe the terahertz electric field [23]. In an isotropic crystal, the refractive index is the same for all crystal axes, and normally, the terahertz-induced refractive index change is small compared to the absolute value of the index, reducing the reflection from the crystal interface.

9.3 Terahertz Time-Domain Spectroscopy

As we have described in the previous sections, coherent terahertz electromagnetic waves can be generated using ultrashort pulsed lasers and can be detected with high sensitivity. In order to perform **THz-TDS** using these methods, the output of the laser is divided into two parts, and one is used for generation and the other for detection of the terahertz wave. An optical delay stage is placed in one of the optical paths to adjust the timing between pump and probe beams. By scanning the time delay, the electric field at different timings can be observed reflecting the actual waveform of the terahertz wave. By means of Fourier transformation, we can obtain the spectrum of the terahertz wave. The signal from the detector is amplified and sent to a computer via an analog-to-digital converter. A chopper is used to switch the terahertz pulse for lock-in detection of the signal. Recently, it has been demonstrated that higher signal to noise ratio can be obtained by scanning the time delay rapidly rather than performing lock-in detection [24, 25]. A typical experimental setup is shown in Fig. 9.2.

An important feature of the THz-TDS is that the electric field is detected instead of the intensity of the terahertz wave. This means that we can observe the phase of each frequency component. Conventional spectroscopy, such as Fourier-transform Infrared (FTIR) spectroscopy, can only measure the intensity, and therefore, the complex reflectivity or transmittivity can only be obtained after **Kramers-Kronig transformation (K-K transformation)**. In the K-K transformation, the data at all frequencies should be known in principle. However, in practical cases, not all of the data can be known precisely enough to calculate the K-K transformation. This results in ambiguity in calculated complex dielectric constants or refractive indices. On the contrary, by using THz-TDS, the complex transmittance or reflectance can be

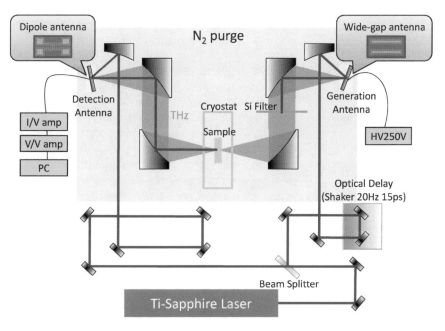

Fig. 9.2 Experimental setup for broadband THz-TDS [25]. An ultra-short pulsed laser with a pulse duration of 15 fs was used with photo-conducting antennas in reflection geometry for the generation and detection of the broadband terahertz wave. The shaker changes the optical delay time of the generated terahertz wave for 15 ps at the oscillatory frequency of 20 Hz. The current induced by the terahertz electric field and the position data of the shaker were simultaneously recorded with a computer and the terahertz waveform was obtained. The setup is purged with nitrogen gas in order to avoid the absorption of water present in air

directly obtained and the complex reflectivity can be calculated without knowledge of other frequency regions.

Here, let us consider a sample with thickness d and **Complex refractive index** n. The complex transmittance of the sample can be calculated as:

$$t = \frac{4n}{(1+n)^2} e^{i(k-k_0)d}. \tag{9.10}$$

By numerically solving this equation for n using the observed complex transmittance t, the complex refractive index can be obtained. Here, k_0 and k are the wavevectors in air and in the sample, and the refractive index of air is assumed to be 1. Multiple reflection is ignored by assuming that d is thick enough to separate the multiply reflected signal from the direct one. Multiple reflections are delayed due to the longer pathlength, so that we can separate these effects in the case of time-domain spectroscopy. In order to get higher spectral resolution, namely, to get a longer time window, we can of course consider **Multiple reflections** in the equation. In the case of thin films, the infinite sum of the multiple reflections can be calculated as follows.

$$t = \frac{4ne^{ikd}}{(1+n)^2 - (1-n)^2 e^{2ikd}}. \tag{9.11}$$

In the case of films deposited on a substrate, we also need to change the equations to take account of the refractive index dispersion of the substrate.

In both cases, the right-hand side of the equation includes the component e^{ikd}. This directly represents the absorption and phase delay of the material. Here, the real part of kd is basically an increasing function, which is nearly proportional to the frequency. Therefore, better results can be obtained when the phase of the transmittance in the left-hand side is also set to be an increasing function of the frequency. In the case of thin films, the phase does not change so much because d is small, whereas in the case of thick samples, the phase data is adjusted by integer multiples of 2π to make the phase approximately a linear function. In order to do this, the phase should be obtained continuously in the frequency axis, and therefore, it is important to consider this when selecting the generation and detection schemes.

We also note that in the case that the terahertz wave is focused at the sample position, we need to correct the Guoy phase, because the focused light contains many wavevectors [26].

9.4 Broadband Terahertz TDS

9.4.1 Photoconducting Antenna

Here, we describe the method using a photoconducting antenna for broadband terahertz spectroscopy. A photoconducting antenna is easy to handle, and can provide a relatively smooth spectrum. The bandwidth is determined by the pulse duration, antenna geometry and absorption due to the antenna substrate in the cases of both generation and detection of terahertz wave [27, 28]. We used a Ti:sapphire oscillator with the pulse duration of 15 fs, and the antennas were placed in a reflection geometry to avoid absorption due to the substrate. The repetition rate, power and center wavelength of the laser were 80 MHz, 400 mW, and 800 nm, respectively. The photoconducting antenna was deposited on a low-temperature-grown GaAs substrate. The shape of the antenna was a 400-μm-gap large arca strip line antenna for generation, and a dipole antenna with a 5 μm gap for detection. In general, the larger the size of an antenna becomes, the smaller the resonant frequency [29]. However, when the sensitivity at low frequency is too high, it becomes difficult to detect the high frequency part, so we used a dipole antenna, which is known to have higher sensitivity at high frequency, for detection. The strip line antenna was chosen for generation because a strong and broadband terahertz wave can be generated when the laser is focused onto the boundary between the strip line antenna and the semiconductor [21, 30].

Fig. 9.3 Observed terahertz
spectrum using the setup
shown in Fig. 9.2 [25]. The
inset shows the corresponding waveform. The detection
bandwidth is expanded to
20 THz without any spec-
tral features, except for a
small frequency range around
8 THz. Due to the rapid scanning using the shaker, a good
signal-to-noise ratio can be
obtained within 500 s

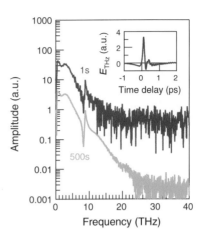

In order to achieve rapid data acquisition and to reduce low frequency electronic
noise and laser fluctuations, we used a 20-Hz shaker with the scanning range of 15 ps
[25]. Figure 9.2 shows the experimental setup. By rapidly scanning the optical delay
using the shaker, observed terahertz waveforms are converted to an electric signal at
$1 \sim 100$ kHz, which results in a reduction of low frequency noise. The observed signal
and the position signal from the shaker are both sent to an analog-digital converter,
and the terahertz electric field is reconstructed by a personal computer.

Figure 9.3 shows the terahertz spectra observed with the setup. Although the
conventional terahertz spectrometer with 100-fs pulsed laser and photoconducting
antenna covers the frequency range up to several terahertz, our setup covers the
frequency region up to 20 THz. In addition, the phonon absorption of the substrate
at 8 THz is considerably reduced thanks to the use of reflection geometry [27]. The
reduced absorption range makes it easier to derive the complex refractive index,
because the phase discontinuity near the absorption is reduced. In addition, as shown
in Fig. 9.3, a good signal-to-noise ratio can be obtained within several minutes by
using a shaker. Therefore, a photoconducting antenna with reflection geometry is
suitable for application to broadband terahertz spectroscopy. Recently it was reported
that similar bandwidth can be realized even in transmission geometry by making the
GaAs substrate thinner by means of etching [31].

9.4.2 Electro-Optic Crystal

Here, we will describe the further broadening of the bandwidth of terahertz spectroscopy by using **Optical rectification**. Since optical rectification does not normally involve real excitation of the material (non-resonant), the photoresponse of the
nonlinear crystal is faster than in the case of the photoconducting antenna. This results
in an increase in the generation bandwidth. The most efficient frequency region for

generation and detection becomes slightly higher than that using the photoconducting antenna, even with the same laser. Taking advantage of this characteristic, we generated an extremely high frequency electromagnetic wave using optical rectification, and tried to detect the electric field by using a photoconducting antenna.

The ultrabroadband detection of terahertz electric fields up to 100 THz has already been demonstrated using an ultrashort pulse laser with 15-fs pulse duration. A GaSe crystal with the thickness of 30 mm is used for generation and a photoconducting antenna for detection [32]. C. Kübler et al. demonstrated the detection of 120 THz fields using a thin GaSe crystal for detection [33]. It is an interesting challenge to extend this bandwidth higher, because it may become possible to reach the **Telecommunication wavelength** of about 200 THz. For this purpose, we used an ultrashort pulse laser with the pulse duration of 5 fs with a **DAST** (4-dimethylamino-N-methyl-4-stilbazolium tosylate) single crystal for generation of the ultrahigh frequency terahertz wave. DAST crystals are known to have a very high nonlinear coefficient [34].

First, we examined the bandwidth of the generated terahertz waves from the DAST single crystal by using normal intensity detectors, such as HgCdTe, and GaAs detectors, coupled with an infrared spectrometer. The observed spectra indicate that if the excitation laser is negatively chirped, a strong terahertz wave up to 200 THz is generated. The negative dispersion is required since the high and low frequency components of the laser should overlap inside the crystal for ultrahigh frequency generation [35].

We then focused the generated terahertz wave onto the photoconducting antenna and tried to detect the electric field. Figure 9.4 shows the experimental setup. In order to achieve the short pulse duration at the antenna position, we placed a thin BaF_2 crystal in the probe beam-path. The Ti-sapphire laser used has a center wavelength of 800 nm, pulse duration of 5 fs, output power of 200 mW and repetition rate of 78 MHz. For electric field detection, we used a dipole antenna with a 5 μm gap.

Figure 9.5 shows the experimental results obtained using the setup schematically illustrated in Fig. 9.4. In the observed waveform, we can clearly see a high frequency oscillation around 800 fs. This delay of the high-frequency component compared to the low-frequency component originates from the fact that the generated terahertz wave has extremely broad bandwidth, and therefore, the refractive index dispersion between high and low frequencies in the DAST crystal and a Ge filter results in a delay of the arrival times at the detector. Analysis of the Fourier transformation of this electric field profile indicates that electromagnetic fields up to 170 THz can be successfully detected using this setup. The spectrum has many dips, such as 50 and 90 THz, which are due to the many infrared-active modes in the DAST crystal (some of them are due to carbon dioxide and water molecules). However, the spectrum is relatively smooth above 100 THz, and in principle, we can perform time-domain spectroscopy to deduce the complex refractive index in this frequency range. Although such a high frequency is far above the high sensitivity region of the photoconducting antenna (several terahertz), which is comparable to the carrier lifetime (several hundred fs), DAST crystals that can generate high frequencies enable us to demonstrate the capability of the photoconducting antenna for observing ultrahigh frequency. In addition, after sen-

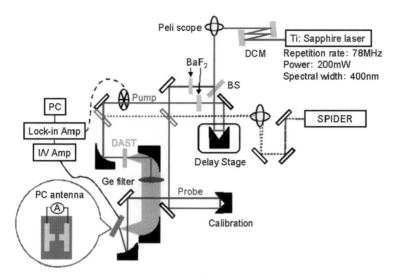

Fig. 9.4 Experimental setup using the 5 fs ultrashort pulsed laser system. The laser output was first negatively chirped using dispersion compensation mirrors (DCM) and the amount of the chirping was controlled by BaF2 plates placed in both the generation and detection arms. The pulse duration was measured by using the **Spectral phase interferometry for direct electric-field reconstruction (SPIDER)** technique. A DAST crystal and photo-conducting antenna were used for generation and detection of the broadband THz wave, respectively. We used the lock-in amplifier and chopper for measurement. A Ge filter was used to cut the low energy tail of the laser itself

Fig. 9.5 (**a**) Waveform and (**b**) spectrum of the ultra-broadband THz wave obtained using the setup shown in Fig. 9.4. The inset shows the horizontal expansion of the waveform around 0.85 ps. The high frequency components correspond to the small oscillations shown in the inset around the delay time of 1 ps

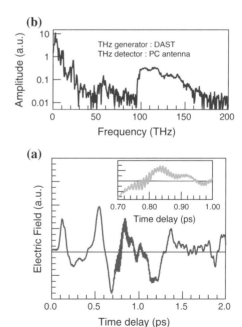

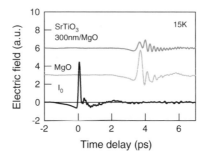

Fig. 9.6 Waveforms transmitted through the MgO substrate, the SrTiO$_3$ (300 nm)/MgO sample and the reference (no sample) [39]. The delay of the arrival time of the terahertz wave is because of the difference of refractive index between the vacuum and MgO. A decrease of the amplitude is observed both in MgO substrate and the SrTiO$_3$ sample

sitivity correction of the detectors and spectrometer, the intensity spectrum of the generated terahertz wave can be obtained. Using the corrected spectrum and comparing it with the Fourier transform of the observed waveform, we can estimate the sensitivity of the photoconducting antenna driven by 5-fs laser pulses. The sensitivity is in good agreement with the theoretical sensitivity obtained from 9.6. It is amazing that this simple 9.6 holds even for the extremely high frequency of 170 THz. From these results, we can conclude that a photoconducting antenna is a suitable detector for a broadband terahertz spectrometer. In the next section, we describe application of the photoconducting antenna to the broadband THz-TDS of materials.

9.5 Applications of Broadband THz-TDS

9.5.1 Characterization of Dielectric Thin Films

As an example of an application of broadband terahertz TDS, we focus on **Ferro-electrics** and related materials, which are important for high-frequency microwave devices, ferroelectric memories, and insulating materials for large-scale integrated circuits. These materials are known to exhibit a smaller dielectric constant as the thickness of the film becomes smaller [36]. The reduction is considered to be due to stress from the substrate, oxygen deficiency, defects or crystallinity. In order to improve the quality of the ferroelectric thin film, it is important to monitor the dielectric constant [37]. Here, we used broadband THz-TDS to characterize dielectric thin films. The terahertz spectroscopy of dielectric thin films has been demonstrated up to 2 THz [38], but we further increased the bandwidth of the spectrometer, which enables us to directly determine the dielectric dispersion of the **Soft mode**. We used the setup shown in Fig. 9.2.

We used a SrTiO₃ thin film deposited on MgO using a pulsed laser deposition technique. SrTiO₃ is a quantum paraelectric material, and has a ferroelectric soft mode that softens as the temperature becomes lower. However, this material does not exhibit any ferroelectric phase transition at low temperature, and maintains an extremely large dielectric constant down to nearly 0 K. However, the dielectric constant depends on the method of thin film fabrication, and is different from that of the bulk single crystal. Figure 9.6 shows the observed time profiles of the terahertz electric field [39]. By means of Fourier transformation, we can convert the waveform to the frequency spectrum, and using the method described in the previous section, we can obtain the complex dielectric dispersion in the terahertz range. In order to get the precise complex transmittance, we deposited the thin film only on a part of the substrate in order to have sample and reference regions on a single substrate.

Figure 9.7 shows the real and imaginary parts of the dielectric dispersion obtained from the experimental results. Both the real and imaginary parts show a strong dependence on the frequency. The peak of the imaginary part is at 2 THz, and there is also a small peak at 5 THz. These peaks are two of the three infrared-active phonon modes that exist in the perovskite lattice structure. The mode with lower frequency (2 THz: soft mode) has a large oscillator strength compared to the other modes, so we analyzed the dispersion using a single Lorentz oscillator model:

$$\epsilon = \epsilon_\infty + \frac{(\epsilon_0 - \epsilon_\infty)\omega_0^2}{\omega_0^2 - \omega^2 + i\gamma\omega}. \tag{9.12}$$

From this fitting, we obtain the resonant frequency and damping constant of the soft mode. However, the frequency is much higher than that in the bulk single crystal (0.2 THz at 4 K). This result is consistent with that of a previous study [36].

In order to reveal the origin of hardening in the SrTiO₃ thin film, we annealed the sample to high temperature and examined the change in the dielectric dispersion

Fig. 9.7 Dielectric dispersions at 15 K of as-grown and high-temperature-annealed SrTiO₃ thin films (460 nm) deposited on MgO [9]. The inset shows the sample configuration. In order to access the precise phase difference due to the thin film, we restrict the area for deposition and use the same substrate for measuring sample and reference data

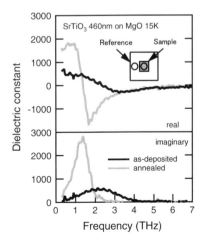

using broadband THz-TDS. The annealing was done at 1200 °C in air for 12 h. The result is also shown in Fig. 9.7. The resonant frequency of the soft mode becomes significantly lower than that in the as-deposited sample. Correspondingly, the low-frequency dielectric constant becomes higher for the annealed sample. These results demonstrate that the film quality strongly affects the soft-mode frequency. Indeed, the atomic force microscope (AFM) image of the sample surface shows an increase in the grain size, and the flat part has become larger after the annealing. X-Ray diffraction measurement also indicates that the lattice constant of the annealed sample becomes closer to that of the bulk crystal [9].

As demonstrated here, broadband THz-TDS can be a powerful tool to investigate the quality of dielectric thin films. It is also possible to discuss the damping constant of the dielectric dispersion. By using this technique for in-situ observation of the film quality during deposition, it may be possible to find a parameter to improve the film quality further, even for thicknesses of the order of several tens of nanometers.

9.5.2 Light-Pump Terahertz Probe Spectroscopy

In this subsection, we review new spectroscopic methods that have become possible after the invention of ultrashort pulse lasers. Especially, light-pump terahertz probe spectroscopy cannot be performed without pulsed broadband terahertz sources. Here, we focus on studies using broadband spectroscopy over 4 THz. The terahertz wave generated with the ultrashort pulsed laser has a pulsed time profile, and therefore, it becomes possible to perform the time-resolved spectroscopy. Using this method, the excited state dynamics of many kinds of materials have been investigated.

R. Huber et al. constructed a broadband terahertz time-domain spectrometer using 30 μm GaSe for generation. They used an ultrashort pulse laser with the wavelength of 800 nm and pulse duration of 10 fs. They performed light-pump terahertz probe spectroscopy to understand the formation mechanism of photoinduced Electron-hole plasma in GaAs [40]. They used 10 μm ZnTe for detection. In this case, a time delay was added to change the timing between the pulses for photoexcitation and those for terahertz electric field detection. Figure 9.8 shows the time dependence of the optical conductivity when the delay time between pump and probe terahertz pulses is scanned. In this figure, we can see the rise time of the plasma response, and simultaneously, the phonon frequency softens due to coupling with the emergence of photoexcited plasma. This result is important because it directly shows how the photoexcited hot carriers form the collective plasma response. As illustrated in this study, the THz-TDS is an important tool for the study of ultrafast responses of materials. Especially, by making the bandwidth broader, we can detect the formation of many quasi-particles.

Exciton formation can also be studied with light-pump terahertz probe spectroscopy. T. Suzuki et al. studied the time-dependence of the photoexcited carriers in Si and revealed their dynamics [41]. They used 30-fs laser pulses from a regenerative amplifier with multipath amplifiers, with 300 μm GaP crystals both for generation

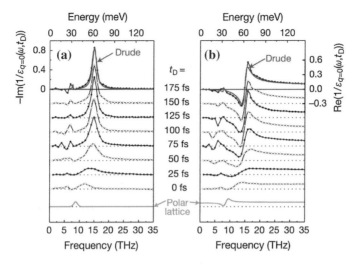

Fig. 9.8 Time dependence of the real (**a**) and imaginary (**b**) parts of the optical loss function ($1/\epsilon$) in GaAs single crystal [40]. The use of thin EO crystals for generation and detection of terahertz waves made it possible to observe the formation of high frequency plasma resonance at 15 THz. The increase of the Drude dispersion after photo excitation can be seen. Reprinted by permission from Macmillan Publishers Ltd., from R. Huber et al., Nature 414 286 (2001)

and detection of the terahertz wave. The bandwidth extended from 0.5 to 6 THz. Figure 9.9 shows the time-dependence of the change in dielectric dispersion as a function of the delay time between pump light and the terahertz probe. Immediately after the photoexcitation, the Drude dispersion due to the photoinduced plasma is observed at low frequency, but as the excitons in Si start to increase, internal transition is observed as the Lorentz dispersion. They observed this kind of time-dependence at several excitation densities and succeeded in revealing the dynamics of a high-density electron-hole system. As demonstrated in this study, the broad bandwidth enables us to discuss the shape of the dielectric dispersion, which is crucial for understanding the physics. High density electron hole systems has been extensively studied to investigate exciton Bose-Einstein condensation, or the phase transition from electron hole plasma to electron hole droplet [42], and THz-TDS could be an important new method to understand the dynamics of these phenomena.

In addition to these studies, light-pump terahertz-probe spectroscopy has been applied to the photoexcited states of superconductors [43], photoinduced response of semiconductor super lattices [44], and other materials to reveal the terahertz dynamics.

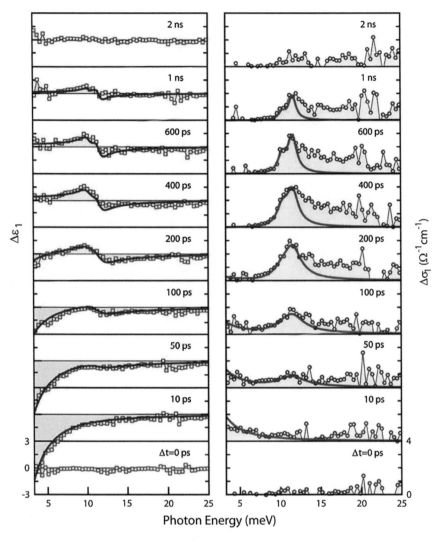

Fig. 9.9 Time dependence of photoinduced change in the dielectric constant (*left*) and optical conductivity (*right*) in a Si single crystal [41]. The conversion dynamics from photoexcited carriers to Excitons can be clearly seen in the THz absorption. Reprinted by permission from the American Physical Society, from T. Suzuki et al., Phys. Rev. Lett. 103 057401 (2009)

9.6 Conclusion

As we have shown in this chapter, broadband THz-TDS can be implemented simply by using an ultrashort pulsed laser and a slight modification of the configuration of generation and detection methods. Furthermore, by combination with photoexcitation using pulsed lasers, we can observe the time-dependence of the dielectric dispersion, which can be used to investigate the physics of various materials, including thin films. Today, many studies are performed in the terahertz region, and the basis of THz-TDS is well established. Therefore, current interest in the field of terahertz spectroscopy is shifting from the development of techniques to the measurement of phenomena which were difficult to access using the previous techniques of far-infrared spectroscopy. New approaches includes the light-pump terahertz-probe spectroscopy, terahertz-pump terahertz-probe spectroscopy [45], and **Nonlinear terahertz spectroscopy** [46–48]. The terahertz region is a new and very promising research area, and broadening the bandwidth of this cutting-edge spectroscopy may reveal intriguing and unexpected physics or chemistry.

Acknowledgments The topics introduced in this chapter are the result of fruitful discussions and collaboration with Hiroshi Shimosato, Ryota Akai, Michitaka Bito, Tadashi Itoh, Dhanvir S. Rana, Iwao Kawayama, Masayoshi Tonouchi at Osaka University.

References

1. C. Rullière (ed.), *Femtosecond laser pulses* (Springer, New York, 2005)
2. M. Chergui, D.M. Jonas, E. Riedle, R.W. Schoenlein, A.J. Taylor, *Ultrafast Phenomena XVII* (Oxford University Press, New York, 2011)
3. D.H. Auston, K.P. Cheung, P.R. Smith, Appl. Phys. Lett. **45**, 284 (1984)
4. M. Tonouchi 2007, Nat. Photon. **1**, 97 (2007)
5. B. Ferguson, X.C. Zhang, Nat. Mater. **1**, 26 (2002)
6. K. Sakai (ed.), *Terahertz Optoelectronics* (Springer, New York, 2005)
7. C. Fattinger, D. Grischkowsky, Appl. Phys. Lett. **54**, 490 (1989)
8. X.C. Zhang, B.B. Hu, J.T. Darrow, D.H. Auston, Appl. Phys. Lett. **56**, 1011 (1990)
9. I. Katayama, H. Shimosato, D.S. Rana, I. Kawayama, M. Tonouchi, M. Ashida, Appl. Phys. Lett. **93**, 132903 (2008)
10. Y. Takahashi, N. Kida, Y. Yamasaki, J. Fujioka, T. Arima, R. Shimano, S. Miyahara, M. Mochizuki, N. Furukawa, Y. Tokura, Phys. Rev. Lett. **101**, 187201 (2008)
11. E. Castro-Camus, M.B. Johnston, Chem. Phys. Lett. **455**, 289 (2008)
12. B.B. Hu, M.C. Nuss, Opt. Lett. **20**, 1716 (1995)
13. J.H. Son, J. App. Phys. **105**, 102033 (2009)
14. Y.S. Lee, *Principles of Terahertz Science and Technology* (Springer, New York, 2009)
15. D.J. Cook, R.M. Hochstrasser, Opt. Lett. **25**, 1210 (2000)
16. J. Dai, X. Xie, X.C. Zhang, Phys. Rev. Lett. **97**, 103903 (2006)
17. I.C. Ho, G. Xiaoyu, X.C. Zhang, Opt. Exp. **18**, 2872 (2010)
18. B. Clough, J. Liu, X.C. Zhang, Opt. Lett. **35**, 3544 (2010)
19. Y. Chen, S. Williamson, T. Brock, F.W. Smith, A.R. Calawa, Appl. Phys. Lett. **59**, 1984 (1991)
20. D.R. Dykaar, B.I. Greene, J.F. Federici, A.F.J. Levi, L.N. Pfeiffer, R.F. Kopf, Appl. Phys. Lett. **59**, 262 (1991)
21. S.E. Ralph, D. Grischkowsky, Appl. Phys. Lett. **60**, 106447 (1992)
22. A. Syouji, S. Saito, K. Sakai, M. Nagai, K. Tanaka, H. Ohtake, T. Bessho, T. Sugiura, T. Hirosumi, M. Yoshida, J. Opt. Soc. Am. **24**, 2007 (2007)
23. Y.C. Shen, P.C. Upadhya, E.H. Linfield, H.E. Beere, A.G. Davies, Appl. Phys. Lett. **83**, 3117 (2003)
24. T. Yasui, Y. Kabetani, E. Saneyoshi, S. Yokoyama, T. Araki, Appl. Phys. Lett. **88**, 241104 (2006)

25. H. Shimosato, T. Katashima, S. Saito, M. Ashida, T. Itoh, K. Sakai, Phys. Stat. Sol. c 3, 3484 (2006)
26. P. Kužel, H. Němec, F. Kadlec, C. Kadlec, Opt. Exp. **18**, 15338 (2010)
27. S. Kono, M. Tani, P. Gu, K. Sakai, Appl. Phys. Lett. **77**, 4104 (2000)
28. S. Kono, M. Tani, K. Sakai, Appl. Phys. Lett. **79**, 898 (2001)
29. M. Tani, S. Matsuura, K. Sakai, S. Nakashima, Appl. Opt. **36**, 7853 (1997)
30. N. Katzenellenbogen, D. Grischlowsky, Appl. Phys. Lett. **58**, 222 (1990)
31. S. Kasai, T. Katagiri, J. Takayanagi, K. Kawase, T. Ouchi, Appl. Phys. Lett. **94**, 113505 (2009)
32. M. Ashida, Jpn. J. Appl. Phys. **47**, 8221 (2008)
33. C. Kübler, R. Huber, S. Tübel, A. Leitenstorfer, Appl. Phys. Lett. **85**, 3360 (2004)
34. F. Pan, G. Knöpfle, C. Bosshard, S. Follonier, R. Spreiter, M.S. Wong, P. Günter, Appl. Phys. Lett. **69**, 13 (1996)
35. I. Katayama, R. Akai, M. Bito, H. Shimosato, K. Miyamoto, H. Ito, M. Ashida, Appl. Phys. Lett. **97**, 021105 (2010)
36. A.A. Sirenko, C. Bernhard, A. Golnik, A.M. Clark, J. Hao, W. Si, X.X. Xi, Nature **404**, 373 (2000)
37. D. Su, T. Yamada, V.O. Sherman, A.K. Tagantsev, P. Muralt, N. Setter, J. Appl. Phys. **101**, 064102 (2007)
38. M. Misra, K. Kotani, I. Kawayama, H. Murakami, M. Tonouchi, Appl. Phys. Lett. **87**, 182909 (2005)
39. I. Katayama, H. Shimosato, M. Ashida, I. Kawayama, M. Tonouchi, T. Itoh, J. Lumin. **128**, 998 (2008)
40. R. Huber, F. Tauser, A. Brodschelm, M. Bichler, G. Abstreiter, A. Leitenstorfer, Nature **414**, 286 (2001)
41. T. Suzuki, R. Shimano, Phys. Rev. Lett. **103**, 057401 (2009)
42. J. Kasprzak, M. Richard, S. Kundermann, A. Baas, P. Jeambrun, J.M.J. Keeling, F.M. Marchetti, M.H. Szymańska, R. André, J.L. Staehli, V. Savona, P.B. Littlewood, B. Deveaud, L.S. Dang, Nature **443**, 409 (2006)
43. R.A. Kaindl, M.A. Carnahan, D.S. Chemla, S. Oh, J.N. Eckstein, Phys. Rev. B **72**, 060510 (2005)
44. R.A. Kaindl, D. Hägele, M.A. Carnahan, D.S. Chemla, Phys. Rev. B **79**, 045320 (2009)
45. M.C. Hoffmann, J. Hebling, H.Y. Hwang, K.L. Yeh, K.A. Nelson, J. Opt. Soc. Am. B **26**, A29 (2009)
46. M. Jewariya, M. Nagai, K. Tanaka, J. Opt. Soc. Am. B **26**, A101 (2009)
47. M. Jewariya, M. Nagai, K. Tanaka, Phys. Rev. Lett. **105**, 203003 (2010)
48. T. Kampfrath, A. Sell, G. Klatt, A. Pashkin, S. Mährlein, T. Dekorsy, M. Wolf, M. Fiebig, A. Leitenstorfer, R. Huber, Nat. Photon. **5**, 31 (2011)

Part III
THz-Technology

Chapter 10
Terahertz Light Source Based on Synchrotron Radiation

Masahiro Katoh

10.1 Introduction

Synchrotron radiation is an electromagnetic wave emitted by relativistic electrons traveling in a strong magnetic field. It provides intense, highly collimated, highly polarized, pulsed, and extremely broadband light source, which is widely used in various research fields especially in the X-ray range. It can also be used in the infrared and terahertz range because it also has a strong spectral component in these regions. In this paper, I Introduce a new way to enhance the infrared and terahertz component of the radiation by far to several orders in the magnitude, using coherent synchrotron radiation. New light source technologies based on the coherent synchrotron radiation are being developed at UVSOR at Institute of Molecular Science, Japan, which will be powerful tools for terahertz applications. Using this light source, we can also utilize the coherence of the radiation for spectroscopy, which was difficult in the studies using normal synchrotron radiation. Although many applications are in the X-ray range, SR is also widely used in a longer wavelength range, such as vacuum ultraviolet (VUV), infrared, or terahertz. In this article, generation of terahertz radiation by synchrotron light sources is reviewed first. Then, coherent synchrotron radiation (CSR), which is much more powerful than the normal SR, is described, focusing on recent developments.

M. Katoh (✉)
UVSOR facility, Institute for Molecular Science, School of PhysicalSciences, Myodaiji-cho, Japan
e-mail: mkatoh@ims.ac.jp

M. Katoh
The Graduate University for Advanced Studies (Sokendai),
Nishigo-naka 38, Myodaiji-cho, Okazaki 444-8585, Japan

K. Shudo et al. (eds.), *Frontiers in Optical Methods*,
Springer Series in Optical Sciences 180, DOI: 10.1007/978-3-642-40594-5_10,
© Springer-Verlag Berlin Heidelberg 2014

Fig. 10.1 Principle of SR.
A relativistic electron is
deflected by a magnetic field
and emits synchrotron radiation in the forward direction

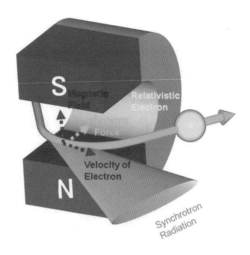

10.2 What is a Synchrotron Radiation Source?

SR is an electromagnetic wave emitted by high-energy electrons traveling in a strong magnetic field (Fig. 10.1). The electrons are deflected by the lorentz force in the magnetic field. This acceleration causes the emission of intense radiation. Electron storage rings are usually used as synchrotron light sources. In these rings, a high-energy electron beam (in the GeV range) circulates for many hours while keeping its energy constant and emitting SR. A synchrotron light source, UVSOR-II, at the Institute for Molecular Science in Japan is shown as an example in Fig. 10.2 [1]. It has 16 light ports around the 50 m ring. Majority of the light ports are used to extract VUV and soft X-rays but one is to extract infrared and terahertz radiation.

Fig. 10.2 UVSOR-II electron storage ring and synchrotron radiation beamlines at the Institute for Molecular Science. The circumference is 53 m. A total of 14 beamlines are operational

In an electron storage ring, an electron beam is deflected along a circular path using a strong magnetic field and SR is generated at the same time. The strength of the magnetic field is usually around 1 Tesla (10 kGauss). Since a synchrotron light source has many light ports, usually much more than ten, it is more cost-effective than might be imagined because of its large scale.

An electron storage ring consists of **bending magnets** (dipole magnets) for deflecting the electron beam, **quadrupole magnets** for focusing the beam, a radio-frequency cavity for supplying the energy lost due to SR, and vacuum system for keeping the electron orbit under an ultra-low pressure, typically 10^{-7} to 10^{-8} Pa. As described previously, the bending magnets act as a radiator. SR emitted by the bending magnets is extracted using a device called a beamline and transported to an experimental apparatus. The beamlines are usually equipped with optical elements such as mirrors or a spectrometer.

Insertion devices are sometimes used instead of bending magnets to produce SR with various properties such as higher brightness, various polarizations, or shorter wavelength. An **undulator** is a device that produces a periodic magnetic field in the electron orbit and makes the electron beam undulate. SR emitted by this device is more bright and quasi-monochromatic. Undulators are mainly used to produce short-wavelength radiation from VUV to X-rays. It is technically difficult to produce infrared or terahertz light by using undulators in an electron storage ring operating in the GeV energy range.

The source size of the SR is determined by the cross section of the electron beam, the diameter of which is typically around 100 microns horizontally (in the orbital plane) and 10 microns vertically (perpendicular to the orbital plane). This small source size indicates that SR is diffraction-limited in the terahertz range.

The temporal structure of the SR is determined by the temporal structure of the electron beam. The electron beam circulating in an electron storage ring forms bunches several tens of picoseconds long due to the radio-frequency acceleration with a repetition rate typically in the range 100–500 MHz.

An electron storage ring does not have the capability of producing an electron beam itself. It stores an electron beam produced by other accelerators such as a linear accelerator or synchrotron. The stored electron beam is gradually lost by scattering with residual gas molecules or with other electrons in the beam. The decay time of the electron beam is called **beam lifetime**, which is typically several hours or several tens of hours. Thus, the electron storage ring should be refilled every several hours or several tens of hours. Recently, a special operation called **top-up injection** was widely introduced in which the electron beam is injected at very short intervals, such as every minute, and the beam intensity is maintained almost constant.

10.3 Characteristics of Synchrotron Radiation

Most of the SR characteristics arise from the fact that it is emitted by ultra-relativistic electrons [2]. Generally, the electromagnetic wave emitted by such an electron is confined within a narrow cone whose opening angle is almost equal to the inverse

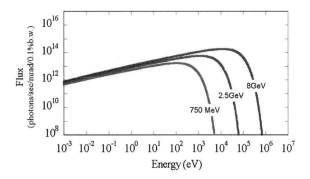

Fig. 10.3 SR spectra for three different values of the electron energy, 750 MeV, 2.5 GeV and 8 GeV, and the same values for the magnetic field strength, 1 T, and the beam intensity, 500 mA. The three values for the energy correspond to those of UVSOR-II, KEK Photon Factory, and SPring-8, respectively

of the **Lorentz factor** (the energy of the electron divided by its rest mass energy). In synchrotron light sources, the energy of the electrons is in the GeV range. Thus, the Lorentz factor is greater than 1,000, which corresponds to an opening angle narrower than 1 mrad. However, it should be noted that the radiation opening angle depends on the radiation wavelength as it is proportional to the one-third power of the wavelength. Thus, the infrared SR extracted by beamlines has a greater opening angle compared with those for shorter wavelengths.

An ultra-wide spectral range is another major characteristic of SR, as shown in Fig. 10.3. In the shorter wavelength range, the spectral range is limited by a parameter called **critical energy**, which is given by the expression [2];

$$\epsilon_c[\text{keV}] = 0.655 E_e^2[\text{GeV}]B[\text{T}] \tag{10.1}$$

The critical energy is proportional to the square of the electron energy for a given magnetic field. This indicates that a high-energy electron storage ring, which is usually large in scale, is necessary to produce hard X-rays. On the contrary, to produce infrared or terahertz radiation, low energy and a relatively small electron storage ring is adequate.

10.4 CSR

10.4.1 What is CSR?

SR emitted by an ensemble of electrons is usually not in phase, as shown in Fig. 10.4 (left). In this sense, SR is not "coherent". How can we make SR coherent?

We have to get light waves emitted by electrons in phase. The simplest way is to confine the electrons within a space shorter than the radiation wavelength. In this

Fig. 10.4 Illustration of a normal synchrotron, CSR from an ultra-short electron pulse, and CSR from a micro-bunched electron pulse

case, all the light waves emitted by the electrons are in phase and the radiation field comes to be coherent. In addition, the radiation field strength is linearly summed and is proportional to the number of electrons. Thus, the radiation power is proportional to the square of the number of electrons.

Generally, the radiation power of an electron pulse can be expressed as [3];

$$P(\omega) = [N + N(N - 1)f(\omega)]P_0(\omega) \tag{10.2}$$

where N is the number of electrons in the pulse, $P_0(\omega)$ is the radiation power emitted by a single electron, and ω is the angular frequency of the radiation. $f(\omega)$ is a **form factor** given by

$$F(\omega) = |\int \exp(i\omega z/c)S(z)dz|^2 \tag{10.3}$$

where $S(z)$ is the longitudinal density distribution of the electron pulse and z is the longitudinal position in the pulse.

The first term on the right-hand side of 10.2 represents normal SR, which is proportional to N. The second term represents **CSR**, which is proportional to the square of N. The second term has a finite value only when $f(\omega)$ is not zero. $f(\omega)$ is the square of the Fourier transform of the electron pulse shape. When the electron pulse length is shorter than the radiation wavelength, the form factor is unity. Since the typical value of N in electron storage rings is greater than 10^{10}, CSR can be much more intense than normal SR by many orders of magnitude, as shown in Fig. 10.5. An interesting case is that in a long electron pulse having a periodic density structure as shown in Fig. 10.4 (right), the form factor has a non-zero value at a wavelength equal to the density structure period and one can observe monochromatic CSR at this wavelength.

10.4.2 CSR from a Linear Accelerator

It is practically impossible to produce electron pulses shorter than the radiation wavelength in the visible range or shorter. However, it is possible to do so in the terahertz or millimeter-wave range. CSR was theoretically investigated in the 1950s [4]. However, it was first observed in the late 1980s [5]. When researchers observed SR emitted by short electron bunches about 1 picosecond long, its intensity was proportional to the square of the intensity of the electron pulses in the millimeter-wave range.

Since this observation, CSR has been produced in many accelerator facilities for applications or for beam diagnostics. In these years, a new concept of a linear accelerator was proposed, called the energy recovery linac. This new accelerator has characteristics of an electron storage ring and a linear accelerator. Ultra-short electron pulses in the sub-picosecond range can be generated with a high repetition rate. Ultra-intense terahertz radiation in the kW range is expected [6].

10.4.3 CSR from an Electron Storage Ring

The electron pulse width in the rings is typically a few tens of picoseconds. Thus, CSR may be emitted in the microwave range. On the other hand, the electromagnetic radiation in a space bounded by metal is suppressed in the long wavelength range such as microwaves [7]. For these reasons, it had been widely believed that CSR could not be observed in electron storage rings. However, several observations of CSR in electron storage rings were reported in the 2000s.

The electron pulse width in an electron storage ring originates from the quantum fluctuation of the SR. The number of photons emitted by a single electron during one circulation in the ring is not large, typically on the order of 10 or 100. The number of emitted photons fluctuates and causes an energy spread in the electron pulse. The electrons having different values of energy travel in different orbits. This results in different orbit lengths. As the speed of the electrons is almost equal to the speed

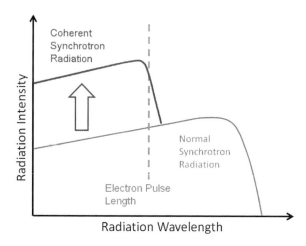

Fig. 10.5 Illustration of a CSR spectrum. In the wavelength range longer than the electron pulse length, coherent radiation takes place, whose intensity is greater than the normal radiation by many orders of magnitudes

of light, the difference in orbit length produces a longitudinal distribution of the electrons and, thus, the electron pulse length. The length is typically a few tens or hundreds of picoseconds.

One cannot control the quantum fluctuation, but one can control the orbital length differences depending on the electron energy. When the electron storage ring is operated under a special condition called low alpha mode, it is possible to suppress these differences and make the electron pulse width short. It was reported that in a few electron storage rings, the electron pulse width could be shortened to a few picoseconds or even shorter and CSR was successfully observed [8]. However, when one increased the electron beam intensity, it was difficult to keep the short pulse width against the strong electromagnetic interaction between the electron beam and the surrounding metal chamber. In reality, CSR was successfully produced with a very small electron beam intensity, typically less than one hundredth of the normal operating condition. However, the CSR intensity is still 10^{2-3} times larger than normal SR with normal electron beam intensity [9]. Stimulated by these results, there are a few proposals to construct small electron storage rings dedicated to CSR [10–12].

On the other hand, the production of CSR from long electron bunches is also being investigated. Electron bunches longer than radiation wavelength can emit CSR when they have a density structure comparable to the radiation wavelength, as shown in 10.2. Laser slicing is a technology to produce a micro-density structure in an electron bunch. Originally it was proposed to slice out a small fraction of the electron pulse and produce a short X-ray pulse [13]. In the electron bunch, a dip structure is created at the same time. It was expected that such an electron pulse with a dip could emit CSR in the wavelength range longer than the dip width [14]. It was successfully demonstrated for several electron storage rings, including the UVSOR-II electron storage ring at the Institute for Molecular Sciences [15, 16].

The experimental setup of UVSOR-II is shown in Fig. 10.6. Laser pulses synchronized with the electron pulses are injected into the electron storage ring and interact with the electron pulses in an undulator. Under a resonant condition between the wiggling motion of the electron beam in the undulator and the oscillation of the laser electric field, a strong interaction takes place and an energy modulation is created in the electron pulse. If the laser pulse is much shorter than the electron pulse, only a small part of the electron pulse is energy modulated. As the electron pulse proceeds in the ring, the energy-modulated electrons escape from their original position and a dip structure is created. The CSR from such a bunch was observed at an infrared/terahertz beamline located at the second bending magnet from the undulator section [17].

As the electron beam intensity increased, the radiation intensity increased proportional to the square of the electron beam intensity, as shown in Fig. 10.7. This indicates the radiation was CSR. As the laser pulse width changed, the spectral range of CSR also changed (Fig. 10.8). The observed CSR intensity per pulse was greater than that of normal SR by 10^{4-5}.

This technique could be introduced in many electron storage rings by utilizing a small part of them. The radiation intensity per pulse was large but the repetition rate was limited by the laser system. The micro-density structure of the electron pulse disappears within a few revolutions. Increasing the repetition rate and average power is the target for our future research. The major advantage of this technology is that by controlling the laser pulse shape, one could control the CSR properties. In Fig. 10.8, we have already shown an example that we can control the CSR spectrum. When we inject laser pulses with periodic amplitude modulation, we can produce tunable and monochromatic CSR from bending magnets, as shown in Fig. 10.9 [18]. By using such a technology, one can control the amplitude, phase, or wavelength of the SR.

10.5 Summary

The infrared/terahertz range of SR has been widely produced in many synchrotron light sources. In recent years, by using several state-of-the-art technologies, production of CSR was successfully demonstrated. Developments in intense terahertz sources based on linear accelerators, especially energy recovery linacs; unique terahertz sources based on electron storage rings in low alpha mode; or the laser slicing/modulation technique and their applications to various research fields would be the subject of future research.

Acknowledgments A part of this work was supported by a Grant-in-Aid for Scientific Research (No. B20360041) from the JSPS, and a grant from the Quantum Beam Technology Program of the Ministry of Education, Culture, Sports, Science and Technology, Japan.

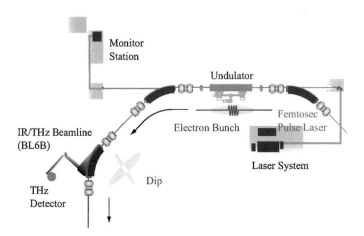

Fig. 10.6 Experimental setup of the CSR using the laser modulation technique. A part of the electron storage ring is illustrated. The figure is taken from [16] under permission of JJAP

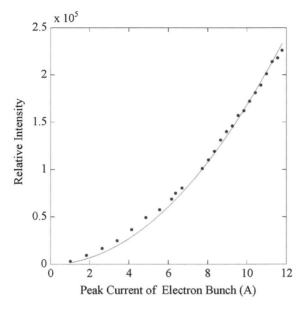

Fig. 10.7 CSR intensity measured at UVSOR-II as a function of the electron pulse intensity. The figure is taken from [16] under permission of JJAP

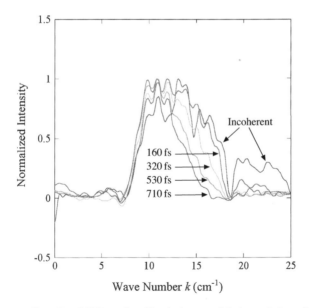

Fig. 10.8 Spectra of broadband CSR produced by the laser modulation technique. It can be clearly seen that the *spectral shape* depends on the laser pulse width. For comparison, all the spectra were normalized to have the same peak intensity. The low energy cut-off is due to the beamline transmission. The figure is taken from [16] under permission of JJAP

Fig. 10.9 Spectra of narrow-band CSR produced by the laser modulation technique for two different values of amplitude modulation period of the laser

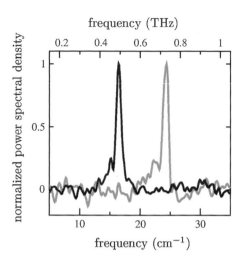

References

1. AIP Conference Proceedings, vol 879, 2007
2. K.J. Kim, *AIP Conference Proceedings*, vol 184, 1989
3. C.J. Hirschmugl, M. Sagurton, G.P. Williams, Phys. Rev. A **44**, 1316 (1991)
4. J. S. Nodvic, D. S. Saxson, **96**, 180 (1954)
5. T. Nakazato, M. Oyamada, N. Niimura, S. Urasawa, O. Konno, A. Kagaya, R. Kato, T. Kamiyama, Y. Torizuka, T. Nanba, Y. Kondo, Y. Shibata, K. Ishii, T. Ohsaka, M. Ikezawa, Phys. Rev. Lett. **63**, 1245 (1989)
6. G. L. Carr, M. C. Martin, W. R. McKinney, K. Jordan, G. R. Neil, G. P. Williams. Nature **420**, 153 (2002)
7. R. Kato, T. Nakazato, M. Oyamada, S. Urasawa, T. Yamakawa, M. Yoshioka, Phys. Rev. E **57**, 3454-3460 (1998)
8. M. Abo-Bakr, J. Feikes, K. Holldack, P. Kuske, W. B. Peatman, U. Schade, G.Wustefeld. Phys. Rev. Lett. **89**, 224801 (2002)
9. M. Abo-Bakr, J. Feikes, K. Holldack, P. Kuske, W.B. Peatman, U. Schade, G. Wuestefeld, H.-W. Huebers, Phys. Rev. Lett. **90**, 094801 (2003)
10. J.M. Byrd, M.C. Martin, W.R. McKinney, D.V. Munson, H. Nishimura, D.S. Robin, F. Sanni-bale, R.D. Schlueter, W.G. Thur, J.Y. Jung, W. Wan, Infrared Phys. Technol. **45**(5-6), 325–330 (2004)
11. H. Hama, H. Tanaka, N. Kumagai, M. Kawai, F. Hinode, T. Muto, K. Nanbu, T. Tanaka, K. Kasamsook, K. Akiyama, M. Yasuda, New J. Phys. **8**, 292 (2006)
12. F. Sannibale, A. Marcelli, P. Innocenzi. J. Synchrotron Radiat., **15**, 655–659 (2008)
13. A.A. Zholents, M.S. Zolotorev, Phys. Rev. Lett. **76**, 912 (1996)
14. R.W. Schoenlein, S. Chattopadhyay, H.H.W. Chong, T.E. Glover, P.A. Heimann, W.P. Leemans, C.V. Shank, A. Zholents, M. Zolotorev, Appl. Phys. B **71**, 1 (2000)
15. K. Holldack, S. Khan, R. Mizner, T. Quast, Phys. Rev. Lett. **96**, 054801(2006)
16. M. Shimada, M. Katoh, S. Kimura, A. Mochihashi, M. Hosaka, Y. Takashima, T. Hara, T. Takahashi, Jpn. J. Appl. Phys. **46**, 7939 (2007)
17. S. Kimura, E. Nakamura, T. Nishi, Y. Sakurai, K. Hayashi, J. Yamazaki, M. Katoh, Infrared Phys. Technol. **49**, 147 (2006)
18. S. Bielawski, C. Evain, T. Hara, M. Hosaka, M. Katoh, S. Kimura, A. Mochihashi, M. Shimada, C. Szwaj, T. Takahashi, Y. Takashima, Nat. Phys. **4**, 390 (2008)

Chapter 11
Single-Photon Counting and Passive Microscopy of Terahertz Radiation

Kenji Ikushima

11.1 Introduction

The terahertz frequency band (1 THz $= 10^{12}$ Hz) occupies a wide region in the spectrum between microwaves and visible light, and the generation, transmission, and detection of THz waves have been marked by various technical difficulties. THz transistors are an ideal target in electronics, where a dramatic increase in the speed of transistors is highly desirable for the purposes of high-speed information processing. Most recently, transistors based on an entirely new principle involving the control of ballistic electron conduction at room temperature have been proposed, and their THz operation is expected [1, 2]. In contrast, in spectroscopic research, measurements in the THz frequency band target extremely low photon energies that are lower than the thermal energy at room temperature. It is thus difficult to utilize conventional ideas based on energy gaps in natural atoms and bulk semiconductors for generators and detectors in the THz frequency band. The recent development of quantum cascade lasers utilizing semiconductor artificial lattices [3] and variable-frequency generators based on parametric amplification [4] has provided new light sources operating in the THz region. Furthermore, a novel method, time-domain spectroscopy (TDS) [5], was developed in the United States about 10 years ago. With ultrashort pulse lasers in the visible and near-IR regions, this method provides a unified measurement scheme for the generation, transmission, and detection of THz waves. It has hence allowed for THz spectroscopy to be routinely performed by using commercial instruments. Because of these rapid advances in THz technology, its practical application in nondestructive testing and medical examinations is expected.

On the other hand, apart from its direct practical application, the THz region has long proven its importance for various research disciplines since it includes a multitude of key energy spectra, such as those due to molecular vibration and

K. Ikushima (✉)
Department of Applied Physics, Tokyo University of Agriculture and Technology,
2-24-16 Nakacho, Koganei, Tokyo184-8588, Japan
e-mail: ikushima@cc.tuat.ac.jp

K. Shudo et al. (eds.), *Frontiers in Optical Methods*,
Springer Series in Optical Sciences 180, DOI: 10.1007/978-3-642-40594-5_11,
© Springer-Verlag Berlin Heidelberg 2014

rotation, lattice vibration in solids, superconducting gaps in metals, and impurities in semiconductors. Although each of these elemental technologies is still being developed, if all problems that they are facing are resolved, THz measurements have vast potential in many fields of scientific research. One of the first imaginable benefits would be an extension of the standard "light measurements," which are used for measuring the transmission and reflection coefficients of external light, into the region of longer wavelengths. Although its extension to longer wavelengths is undoubtedly important, it is not the only appealing feature of the THz frequency band. Another important approach that we considered is to capture extremely weak spontaneous THz radiation emitted from measurement objects without external irradiation.

11.2 Challenges in Passive THz Sensing

Approaches to passive measurement without external irradiation have long been sought in the field of astronomy for the purpose of capturing background radiation, which carries information about the initial moments after the creation of the universe. However, considering the abundant characteristic spectra in matter, radiation from microscopic areas in molecules and solids can certainly provide an abundance of intriguing information since the THz photon energy (about 10 meV) is lower than that of the thermal energy at room temperature (25 meV at 300 K) and constitutes only about $1/100$ of that of visible or near-IR light. This implies that various objects can easily obtain the energy necessary for emission of THz waves even when no special conditions have been applied. In other words, target objects can acquire energy of the order of 10 meV from their environment in the form of thermal energy or a chemical process, such as the hydrolysis of ATP. In addition, the acquired additional energy is transformed into a characteristic mode of the target object itself, and part of it is emitted as THz radiation.

Since molecules in living organisms are usually polarized, their vibration and rotation generate THz waves. For example, when molecular motors are activated by consuming ATP in a thermal bath, it might be possible to determine which internal degrees of freedom the molecules are using in their activity through THz photon-counting measurements. In addition, THz radiation would also find extensive application in thermography of a living single cell at slightly more macroscopic scales where temperature can be defined. When a molecule of ATP is consumed, a local increase in temperature occurs because the released energy is distributed to nearby water molecules. The ability to perform real-time observation of the exact places and times of ATP consumption within a group of molecules would provide an unparalleled method for research on metabolic mechanisms. Since strong external light is used in standard THz measurements, the target objects are excited, and such "intrinsic" emission of light cannot be observed. In order to study the microdynamics of biological systems in their natural state, it is desirable to perform passive microscopic observations without irradiation. Moreover, since various semiconducting quantum structures and molecular devices possess spectra in the THz band, we can expect

THz emission with a threshold of about 10 mV for the source-drain voltage. Passive THz microscopes would be useful for researching the mechanism of THz emission from such microscopic devices and developing novel THz devices.

Although thermography is one representative example of **passive imaging**, existing instruments cannot be used for the above-mentioned purposes due to their limitations in terms of available bands and sensitivity. Thermal radiation on a macroscopic scale originates from the dynamics of electrons, ions, and molecules. However, conventional technology is capable of detecting only thermal radiation averaged over space and time. The ability to observe a local nonequilibrium state within a short-time interval would allow for studying the true dynamics of electrons, ions, and molecules on a microscopic level. Since radiation intensity measured with a passive method is generally proportional to the square of the area and is thus rather small, photon-level sensitivity would be necessary if high-spatial resolution is required. A dramatic improvement in sensitivity would entail enhanced spatiotemporal resolution, resulting in the acquisition of qualitatively different information in comparison to conventional thermography. On the other hand, the attainment of spatial resolution which far surpasses the diffraction limit would be possible (if allowed by the detection sensitivity) by using apertureless near-field technology [6–8]. In reality, it has been reported that spatial resolution of 150 nm (for wavelength of 150 μm) has been achieved in active measurements [8]. To work toward the goal of realizing **passive microscopy** of extremely weak THz radiation, we are in the process of developing highly sensitive measurement technology for the THz frequency band. In this paper, we present the current state of THz photon-counting technology.

11.3 Photon Counting of THz Radiation

11.3.1 Sensitivity of Conventional Detectors

Figure 11.1 presents the **specific detectivity** D^* of a conventional THz and IR detector (the operation temperature is shown in brackets). In general, an index for the detection sensitivity is defined through the noise-equivalent power (NEP), for which the power of the incident light achieves a signal-to-noise ratio $S/N = 1$. However, in comparing the sensitivity of different detectors, it is necessary to perform normalization with respect to the area of the photoactive region, A, and the bandwidth (measurement integral time), Δf. Since the signal strength increases proportionally to the area A of the detector and the noise increases proportionally to $\sqrt{A}$, the inverse of NEP normalized to the area is used for comparing the performance of detectors. This is the specific detectivity $D^* = \sqrt{A}\sqrt{\Delta f}/(\text{NEP})$ cmHz$^{1/2}$/W. The S/N ratio is defined as the ratio of the signal current (or voltage) to the noise current (or voltage). However, apart from the intrinsic noise of the detector itself, statistical fluctuation of photon flux in background radiation also causes noise current (or voltage). The continuous line for D^*_{BLIP} (background limited IR performance) in Fig. 11.1 represents the

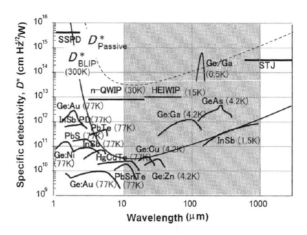

Fig. 11.1 Specific detectivity D^* of conventional IR detectors

performance limit of an IR detector operated at room temperature. Here, we consider the case of a detector with black-body radiation at 300 K, a viewing angle of 2π and a quantum efficiency (signal counts/number of incident photons) $\eta = 1$. Since the passive method mentioned above aims at observing spatiotemporal fluctuations, in a sense, it can be considered that it focuses on precise measurement of photonic noise. Therefore, the specific detectivity required for our passive measurement must be considerably higher than D^*_{BLIP}. In actual microscopes, there are losses in the optical system, and the quantum efficiency of the detector η is less than 1. Therefore, according to rough estimations, the sensitivity should be at least 1,000-fold higher. The detectivity D^*_{Passive} required for passive measurements is shown as a broken line in Fig. 11.1, from which it can be seen that few conventional detectors exceed this level of sensitivity.

11.3.2 Charge-Sensitive Single THz Photon Detectors

Ultrasensitive optical measurements are performed by capturing and counting single photons (the smallest units of radiant energy). photon counters for the visible and near-IR regions have already been implemented by using photon multipliers and avalanche Si diodes. These photon counters are indispensable for measurements of ultralow light level phenomena, such as that required for single molecule detection through fluorescence microscopy [9] and for near-field imaging [6]. The photon-counting method, however, has not been available in the THz region, because the photon energy is lower than the thermal energy at room temperature. Although the spectral response range of superconducting detectors has been expanded toward the longer wavelength region [10–12], a new type of charge-sensitive single-photon detector based on semiconductor quantum structures has been developed in recent

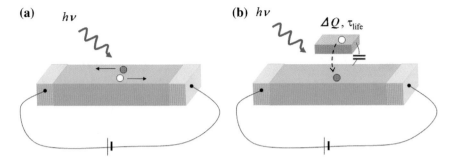

Fig. 11.2 Schematics of conventional photoconductors (**a**) and a new type of THz detector (**b**)

years. The photon detectors have made it possible to perform photon counting across the mid-IR, far-IR, and sub-millimeter wave regions [13–21].

After photoexcitation in conventional detectors, the energy of the absorbed photon is dispersed into the electron system or the lattice system, and a certain state after dispersion is read as an electrical signal. Figure 11.2a shows a schematic illustration of a conventional photoconductive detector. A single-photon absorption event excites a nonequilibrium electron (or hole), resulting in a photocurrent signal. Therefore, even if the efficiency of conversion into a nonequilibrium electron is unity, the upper limit of the photoresponse signal is evaluated as $I_{sig} \leq (P/h\nu)e$ (P, power of the incident photon; $h\nu$, photon energy; e, unit charge). This imposes a limit on the value of D^* of conventional detectors.

A photon is a high-quality packet of energy in a low-entropy state. An approach based on detecting signals after the packets of energy have been dispersed into systems with innumerable degrees of freedom, for example, into a lattice system or many-electron system, is therefore not the most appropriate for detecting individual low-energy photons. A charge-sensitive detection scheme has been devised to transform such a low-energy photon into a single excited electron having a long lifetime without dispersion and subsequently to detect the excited charge electrically [13, 16, 19]. As shown in Fig. 11.2b, the charge detector consists of a transistor with a floating gate. The floating gate is composed of an isolated semiconducting two-dimensional electron gas (2DEG) layer, which serves as a THz wave absorber. The detection mechanism is as follows: (1) an excited electron is ejected from the floating gate when a single photon is absorbed, (2) a long-lived unit charge is created in the gate, and (3) the induced charge is detected by a charge detector (transistor) near the gate. An important point is that the number of electrons detected as the response to a single-photon absorption event is $G \equiv (\Delta I \tau_{life})/e \; (\gg 1)$ if the lifetime of the photo-induced charged state is taken as τ_{life}. Here, ΔI is the amount of current change per photo-induced charge in the charge detector. This implies that an integral function for the lifetime τ_{life} is built into the THz detector itself. As a result, the signal current is amplified to $I_{sig} = G \times (P/h\nu)e$ with a gain of G. Although ΔI and the lifetime τ_{life} depend strongly on the design of the detector, a lifetime of $\tau_{life} = 10^{-8} - 10^3$ s has typically been estimated in earlier work [14] (the corresponding gain is $G = 10^4 - 10^{12}$). This results in a large D^* that is far greater than that for any conventional device, thus allowing for the detection of individual photons even in the THz band.

All charge-sensitive THz detectors discussed here are fabricated from a GaAs/AlGaAs heterostructure crystal. The detectors can be divided into two main groups. One group includes devices using a single-electron transistor (SET) as a charge detector [13–18] (quantum-dot (QD) detectors: $D^* = 10^{17} - 10^{19}$ cmHz$^{1/2}$/W, wavelength $\lambda = 0.1 - 1$ mm), and the other group is based on a photo-transistor with a double-quantum well [19–21] (charge-sensitive IR Phototransistor (CSIP); $D^* = 10^{15}$ cmHz$^{1/2}$/W; wavelength, $\lambda = 0.01 - 0.07$ mm). One of the two quantum wells (the upper quantum well) serves as an isolated floating gate, and the other (the lower quantum well) serves as a conducting channel. The detector can be considered a field-effect transistor with a photoactive semiconducting gate.

In terms of the photon absorption mechanism, QD detectors are further classified into (i) a magnetically tunable single-QD type [13–15] that uses cyclotron resonance in strong magnetic fields (typically $\lambda = 0.1 - 0.2$ mm; a tunable range of about 40 % with a bandwidth of 5 % or less for an individual detector), (ii) a double-QD type that uses plasma resonance in the QD [16] (typically $\lambda = 0.6$ mm), and (iii) a hybrid type that combines a semiconductor QD with an Al-junction SET [17, 18] ($\lambda > 0.2$ mm).

The detectors are fabricated on a GaAs/AlGaAs heterostructure crystal by standard electron beam lithography. Figure 11.3 presents a single QD type detector mounted on an optical system. An aplanatic Si hyperhemispherical lens is used as a substrate lens with its focal point at the planar antenna of the QD detector (Fig. 11.3a–c) [15]. Figure 11.3d shows a scanning electron micrograph of the QD detector. A 2DEG layer is formed at a depth of around 100 nm from the surface of the crystal. When metal gates are negatively biased, the 2DEG layer is fully depleted from the region below the gates and a 2DEG island of 500 nm in diameter (a so-called QD) is formed that has about 300 electrons. A schematic of a QD is shown in Fig. 11.4. When the QD is coupled to electron reservoirs on both sides (source electrode S, drain electrode D) through weak tunnel junctions, the QD serves as an SET. Here, one gate electrodes (G) is used to control the electrostatic potential of the QD. These metal gates, stretched up and down, serve as a planar bow-tie antenna (Fig. 11.5a). The single-QD detector can be effectively operated via the application of a strong magnetic field perpendicular to the plane of the 2DEG layer. In an external magnetic field, B, of the order of a few teslas, the lowest orbital Landau level (LL) is fully occupied by electrons while the first excited LL is partially occupied by a few electrons. As shown in Fig. 11.5b (for simplicity, spin splitting is ignored), these LLs form two compressible metallic regions at the Fermi level, serving as a current path (the outer ring) and a floating gate (the inner core). When a THz photon with energy corresponding to the LL spacing is resonantly absorbed into the QD (cyclotron resonance), an electron-hole pair is created as shown in Fig. 11.5b. The excited electron and hole are quickly separated within the potential barrier and arrive at the inner core and the outer ring, respectively. The QD is thus electrically polarized. Since the electron-hole pair is spatially separated, the photoexcited state has a long lifetime and nonequilibrium electrons accumulate at the floating gate (the inner core). This unit

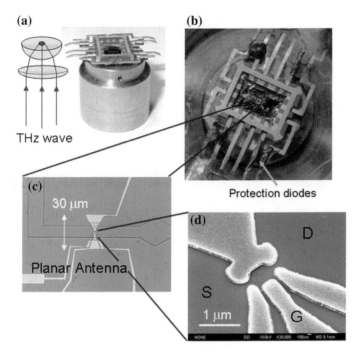

Fig. 11.3 Quantum-dot single THz photon detector mounted on a silicon lens

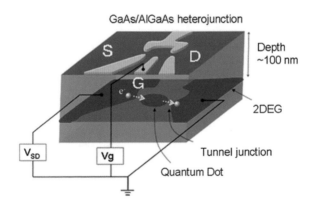

Fig. 11.4 Schematic representation of a semiconductor QD serving as a single-electron transistor

charge of $-e$ induced in the floating gate changes the tunneling current of the SET, resulting in detection of a single-photon absorption event via a conductance switch of the SET. Figure 11.5c shows a typical example of telegraph-like conductance signals due to photon counting.

Fig. 11.5 QD single THz photon detector. **a** Schematics of the detector. Metal gates work as a planar antenna. **b** Energy profile of two-dimensional electrons in a QD under strong magnetic field (*left*). Landau levels form two conducting areas, an outer ring and an inner core of the QD at the Fermi level (*right*). **c** Example of telegraph signals measured by single-photon counting

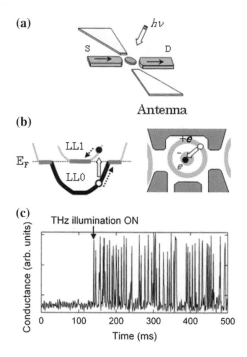

The operation temperature of QD detectors is determined by that of the SET. The highest operation temperature is about 0.4 K in our QDs. If a quantum-point contact [22] or an FET with a microscopic gate [19] is used as the charge detector, it will be possible to raise the limit for the operation temperature to 4 K. In this regard, for wavelengths longer than 0.1 mm, the dark switching rate W_{dark} of the QD detector far exceeds the photon-counting rates arising from thermal radiation in a 6 K environment. Therefore, it is necessary to maintain the temperature of the detector and its environment under 4.2 K regardless of the specific detection mechanism.

The intrinsic limit for the operation speed is determined by the lifetime of excited electrons in the floating gate (the inner core). Since the lifetime of photoexcited states depends strongly on the wave-function overlap, it is highly sensitive to the distance between the core and the ring (or the size of QD). The core-ring spacing can be tuned by the magnetic field and the gate-voltage conditions. Thus, the lifetime can be widely adjusted from a few nanoseconds to more than 10 s. In actual usage, however, the operation speed is restricted by the instrumental time constant for measuring SET currents.

An SET is a fast, low-noise device. The intrinsic limit on the speed of an SET is thought to be of the order of 10 GHz. At present, SET currents are measured through a low-noise preamplifier at room temperature, and thus the operation speed is limited to the order of 10 μs by the time constant attributed to the capacitance of coaxial cables running through the cryostat [15]. Nevertheless, operation speed of the order of 10 ns will be realized when a cryogenic preamplifier placed near the detector is introduced.

The effective sensitivity of a photon counter is determined by the dark switching rate W_{dark} and the quantum efficiency η, which can be expressed in the form of $W/Hz^{1/2}$ [23]. The quantum efficiency η of QD detectors is mainly determined by the coupling between the antenna and the QD. For THz waves entering the planar antenna from vacuum, the efficiency is roughly estimated to be $\eta = 0.1 - 1$ and the measured value of W_{dark} is 0.001 s^{-1} at 70 mK (for a single QD type). With these parameters, we obtain NEP $= 10^{-21}$ W$/Hz^{1/2}$ at 70 mK, which corresponds to a more than 1,000-fold increase in sensitivity in comparison with conventional THz detectors. Since the photoactive area S of the detector is determined by the wavelength λ_ϵ of GaAs medium, the specific detectivity is evaluated to be $D^* = 10^{18}$ cmHz$^{1/2}$/W. Although quantum efficiency is considered to be exceedingly low, the QD detector yields an incomparably higher sensitivity.

11.4 Photon-Counting THz Imaging

11.4.1 Scanning Confocal THz Microscope

We have recently developed a photon-counting THz microscope [15] for use in fundamental research on semiconductors. In this microscope, a QD detector (single QD type) is mounted on a scanning confocal optical system. Schematic and cross-sectional views of the microscope are presented in Fig. 11.6. The QD detector is cooled to 0.3 K with a helium-3 refrigerator, whereas the sample temperature is maintained at 4.2 K by means of a thermal insulating vacuum layer. The dark switching rate W_{dark} increases to about 0.1 s^{-1} under operation conditions of 0.3 K. Nevertheless, in the optical system of this microscope, THz waves are irradiated onto the planar bow-tie antenna of the detector from the side of the dielectric substrate lens, and consequently the quantum efficiency is improved to several tens of percent. Therefore, under the condition that a substrate lens is used, the sensitivity of a QD detector can be estimated as $NEP = 10^{-21}$ W$/Hz^{1/2}$ and $D^* = 10^{18}$ cmHz$^{1/2}$/W even at a temperature of 0.3 K. On the other hand, an aplanatic hyperhemispherical Si lens is also used on the objective lens side, to which the measured sample is attached, and functions as a Si-solid immersion lens [24]. By moving a sample fixed on a piezo-driven XY stage, the focal point of the solid immersion lens traverses the entire surface of the sample. A spatial resolution of 0.05 mm is achieved, which exceeds the diffraction limit for a free-space wavelength of 0.13 mm. It should be noted that even if loss in the optical system is considered, it still functions as a microscope with a sensitivity of 10^{19} W$/Hz^{1/2}$ (detecting about 100 photons/s).

In actual measurements, it is necessary to pay careful attention to both electrical and optical background noise. SETs, which detect individual elementary charges in their environment, are exceedingly sensitive to electrical noise. As shown in Fig. 11.6,

(a)   (b)

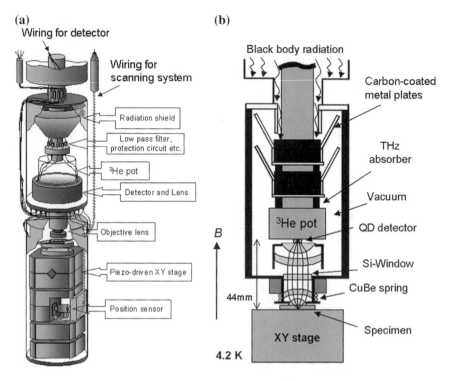

Fig. 11.6 Schematics of photon-counting THz microscope with a QD detector

therefore, all wires except those used for the detector are placed outside the helium-3 refrigerator, and they are shielded by multilayered metallic pipes. Furthermore, thermal radiation straying from higher temperature environments (from the upper part of the microscope) through the thermal insulating vacuum layer is blocked by metal plates covered with a carbon coating. A passive THz microscope capable of performing spectrometry has also been developed which uses a similar optical system [25].

11.4.2 THz Imaging of Quantum Hall Effect Devices

Under strong magnetic fields, the kinetic energy of a 2DEG splits into a series of LLs. In the quantum Hall effect (QHE) states, the longitudinal resistance vanishes and the Hall resistance R_H is quantized. Since the zero longitudinal resistance means that the electric field component parallel to the direction of the current is zero, the 2DEG can be regarded as an energy dissipationless conductor. Therefore, there is neither light emission nor even Joule heating. Nevertheless, the following issue arises: The two-terminal resistance in the QHE states corresponds to the Hall resistance R_H; hence, resistance of R_H can be considered to be connected to the current source and energy dissipation of $R_H I^2$ necessarily occurs somewhere in the conductor. At first glance, one might wonder whether some dissipation occurs near the current contacts. This fundamental issue has been discussed numerous times since the discovery of the QHE, and it is currently understood as follows. First, a schematic representation of the electrostatic potential (with respect to the electrons) in the QHE state is presented in Fig. 11.7a. Singular points (called hot spots) of the potential are formed at two diagonal corners of the current contacts. As seen in Fig. 11.7b, while equipotential lines traverse the central part of specimen along the length of the Hall bar, strong electric fields are locally generated at the two corners of the current contacts. When highly energetic electrons are injected the 2DEG layer through the hot spot of the source corner, most of the excess energy will be released into the lattice system (most probably through acoustic phonon emission). In a similar manner, it can be expected that the excess energy will be dissipated to the lattice system when the two-dimensional electrons are ejected to a metal contact through the hot spot of the drain. However, as can be seen below, when the potential difference between the source and the drain corresponds to the LL energy spacing (cyclotron energy), nonequilibrium electrons are generated in the upper LL in the process of entrance and exit of electrons. Since cyclotron radiation takes place via their relaxation, part of the dissipation energy $(R_H I^2)$ is consumed for emission of THz electromagnetic waves. Figure 11.8 shows an example of photon-counting imaging [15] as obtained with the THz microscope mentioned above. The extremely weak THz radiation emitted from a Hall-bar device fabricated from a GaAs/AlGaAs heterostructure crystal is depicted. This passive THz imaging reveals the spatial distribution of nonequilibrium electron concentration generated in quantum transport phenomena [26].

Fig. 11.7 a Schematics of electrostatic potential of a Hall bar in the quantum Hall effect state. **b** Equipotential lines of the Hall bar

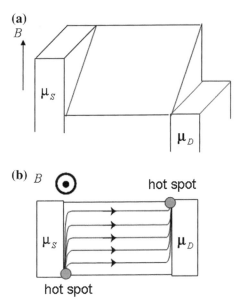

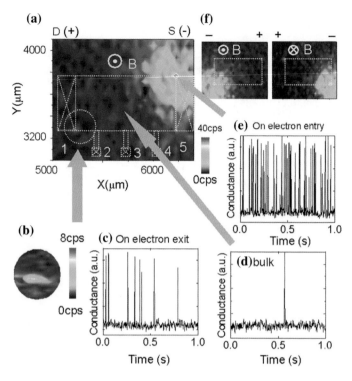

Fig. 11.8 Photon-counting imaging of THz radiation in a quantum Hall effect device. Radiation is found at two corners of current terminals

Let us discuss the mechanism of photon emission shown in Fig. 11.8 in further detail. The sample is subjected to a strong magnetic field ($B = 5.62$ T, cyclotron energy $\hbar\omega_c = 9.43$ meV, which corresponds to a wavelength of 132 µm) and is set into an integer QHE state at the LL filling factor $\nu = 2$. The source-drain voltage applied between the terminals 1 and 5 of the Hall bar is $V_{SD} = V_{15} = 15.5$ mV (the corresponding current is 1.2 µA). As shown in Fig. 11.8, when the source-drain voltage exceeds the cyclotron energy $\hbar\omega_c/e$, cyclotron radiation occurs at the diagonal corners (hot spots) of the current contacts. The power of the radiation emitted at the exit corner is of the order of 0.01 pW (10^4 photons/s) (Fig. 11.8b). In previous studies on cyclotron radiation [27, 28], due to a lack of detector sensitivity, it was necessary either to apply a corresponding voltage far exceeding the cyclotron energy or to arrange more than 10,000 Hall-bar devices. By using quantum-dot single-photon detectors, separate observations can be performed for each current contact (source or drain) as well as to monitor individual Hall-bar devices near the threshold voltage $\hbar\omega_c/e$ for photon emission. Furthermore, the threshold voltage for photon emission is decisively determined (Fig. 11.9a). At the entrance of electrons (source corner), the threshold voltage for emission is $V_{SD} = \hbar\omega_c/(2e) = 4.72$ mV (0.365 µA), and at the exit (drain corner) it is $V_{SD} \simeq \hbar\omega_c/e = 9.43$ mV (0.730 µA). This

Fig. 11.9 **a** Photon-counting rates as a function of source-drain voltages applied to the quantum Hall effect device. Bottom schematics represent energy profile of electrons at the boundary between the metal region and the 2DEG layer (at the electron-entry corner of the source (**b**) and at the exit corner of the drain (**c**))

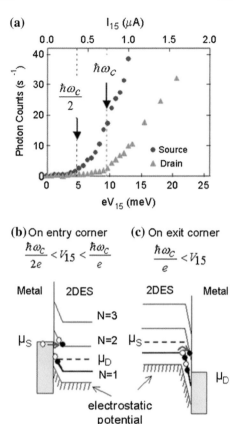

experimental finding is well explained by electron tunneling through the potential barrier (or valley) around the source (or drain) contact (see Fig. 11.9b and c) [15, 28]. In addition, the energy distribution of nonequilibrium electrons can be measured by passive THz spectroscopy [25].

Figure 11.10 shows a measurement example of cyclotron radiation generated by one-dimensional current (edge channel) flowing along the edge of a Hall bar [26, 29]. In this sample, metal gates across the conducting channel are deposited. Here, both the lowest and the upper LLs are fully occupied by electrons in the QHE state at the filling factor $\nu = 4$. Two edge channels are formed along the sample boundaries, namely, an outer edge channel (at the lowest LL) and an inner edge channel (at the upper LL). When a potential barrier is introduced by negatively biasing the cross gate, the inner edge channel is reflected while the outer edge channel is transmitted through the barrier (Fig. 11.10a). Under such conditions, an edge channel injected from a source contact collides with the one injected from a drain contact at the diagonal corners of the cross gate (indicated with circles in Fig. 11.10a). THz radiation can thus be expected when the source-drain voltage reaches the level of

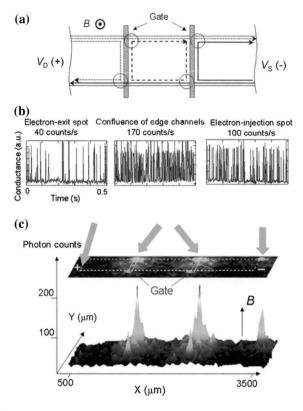

Fig. 11.10 THz photon generation due to electrons injected by edge channels. **a** Design of the sample. **b** Telegraph signals of photon counting. **c** Photon-counting image of cyclotron radiation

cyclotron energy. Figure 11.10c shows the results of photon-counting imaging. As expected, THz radiation is found to be at the confluence of edge channels.

In the QHE regime, edge channels provide well-defined one-dimensional conducting channels, for which electrically controlled beam splitters can be used. The edge channels have been utilized in quantum interference demonstrations such as the fermionic Hanbury-Brown-Twiss experiment [30] and the electronic Mach-Zehnder interferometer [31]. The experimental results shown in Fig. 11.10 indicate that the dissipationless (noiseless) electron beam serves as an injection current for photon generation, implying the possibility of a nonclassical THz generator in which photon number fluctuation would be suppressed beyond the Poisson limit. Furthermore, since this monochromatic THz point source can be electrically controlled through the gate voltage, a unique concept has been proposed for THz-photon circuits, in which both THz point sources and QD detectors are incorporated in a single device and they are optically connected via coplanar transmission lines [32]. This photon circuit is also deemed to be a complete on-chip THz optical system, in which all optical components are mounted on a single device. If all efficiencies for emission, transmission, and detection approach unity, quantum optical measurements would be possible with solid-state devices. Considering the technical challenges in fabrication for optical confinement of short-wavelength visible light and the difficulty of photon counting of microwaves and millimeter waves, THz spectral region might be appropriate for single-photon-level control in solid-state devices.

In addition, although it was not discussed here, the possibility of THz lasing through inverted population in the edge states [33, 34] as well as the self-organization of nonequilibrium electron distribution at large currents [35] has been found. As described above, passive THz measurements can be successfully used to elucidate the nature of electron dynamics in low-dimensional semiconductors, and this sensitive THz measurement technique should enable acquisition of electron information that has so far been unobtainable through electrical resistance measurements, optical measurements, scanning tunneling microscopy, and other conventional measurement methods.

11.5 Future Perspectives

The application of passive THz measurements can be extended to cover broader areas of research, not limited to the fundamental physics of low-temperature semiconductors. If thermal insulating vacuum layers are introduced as described above for microscopes, observations of samples at higher temperatures (even at room temperature) will be possible. Recently, a research group in France has reported an experiment in which thermally induced near-field IR signals are detected on a heated thin film of metal deposited on SiC through a conventional detector (HgCdTe; wavelength, $10 \, \mu m$) [36]. The dramatic improvement in sensitivity by virtue of single-photon detectors will facilitate the further development of passive THz and IR microscopy.

Acknowledgments We thank S. Komiyama, H. Sakuma, T. Hasegawa, Y. Yoshimura, T. Ueda, and K. Hirakawa for helpful contributions to this article. This work is supported by JST PRESTO and the Asahi Glass Foundation.

References

1. Q. Diduck, M. Margala, M.J. Feldman, *Proceedings of the, IEEE MTT-S International Microwave Symposium Digest*, 345 (2006)
2. D. Huo, Q. Yu, D. Wolpert, P. Ampadu, J. Emerg. Technol. Comput. Syst. **5**, 5 (2009)
3. R. Köhler, A. Tredicucci, F. Beltram, H.E. Beere, E.H. Linfield, A.G. Davies, D.A. Ritchie, R.C. Iotti, F. Rossi, Nature **417**, 156 (2002)
4. K. Kawase, H. Minamide, K. Imai, J. Shikata, H. Ito, Appl. Phys. Lett. **80**, 195 (2002)
5. P.Y. Han, X.-C. Zhang, Meas. Sci. Technol. **12**, 1747 (2001)
6. S. Kawata, Y. Inouye, T. Kataoka, T. Okamoto, *Nano-Optics*, eds. by S. Kawata, M. Ohtsu, M. Irie (Springer-verlag, Berlin, 2002) p. 75
7. B. Knoll, F. Keilmann, Nature **399**, 134 (1999)

8. H.-T. Chen, R. Kersting, G.C. Cho, Appl. Phys. Lett. **83**, 3009 (2003)
9. S.A. Soper, M. Wabuyele, C.V. Owens, R.P. Hammer, *Single Molecule Detection in Solution*, eds. by Ch. Zander and J. Enderlein, R. A. Keller (Wiley-VCH, Berlin, 2002) p. 323
10. A. Lipatov, O. Okunev, K. Smirnov, G. Chulkova, A. Korneev, P. Kouminov, G. Gol'tsman, J. Zhang, W. Slysz, A. Verevkin, R. Sobolewski, Supercond. Sci. Technol. **15**, 1689 (2002)
11. S. Bechstein, B. Beckhoff, R. Fliegauf, J. Weser, G. Ulm, Spectrochim. Acta Part B-At. Spectrosc. **59**, 215 (2004)
12. J. Wei, D. Olaya, B.S. Karasik, S.V. Pereverzev, A.V. Sergeev, M.E. Gershenson, Nat. Nanotechnol. **3**, 496 (2008)
13. S. Komiyama, O. Astafiev, V. Antonov, T. Kutsuwa, H. Hirai, Nature **403**, 405 (2000)
14. V. Antonov, O. Astafiev, T. Kutsuwa, H. Hirai, S. Komiyama, Phys. E **6**, 367 (2000)
15. K. Ikushima, Y. Yoshimura, T. Hasegawa, S. Komiyama, T. Ueda, K. Hirakawa, Appl. Phys. Lett. **88**, 152110 (2006)
16. O. Astafiev, S. Komiyama, T. Kutsuwa, V. Antonov, Appl. Phys. Lett. **80**, 4250 (2002)
17. H. Hashiba, V. Antonov, L. Kulik, A. Tzalenchuk, P. Kleinshmid, S. Giblin, S. Komiyama, Phys. Rev. B 73 081310(R) (2006)
18. J.C. Chen, Z. An, T. Ueda, S. Komiyama, K. Hirakawa, V. Antonov, Phys. Rev. B **74**, 045321 (2006)
19. Z. An, J.C. Chen, T. Ueda, S. Komiyama, K. Hirakawa, Appl. Phys. Lett. **86**, 172106 (2005)
20. Z. An, T. Ueda, K. Hirakawa, S. Komiyama, IEEE Trans. Electron Devices **54**, 1776 (2007)
21. T. Ueda, Z. An, K. Hirakawa, S. Komiyama, J. Appl. Phys. **103**, 093109 (2008)
22. S. Pelling, R. Davis, L. Kulik, A. Tzalenchuk, S. Kubatkin, T. Ueda, S. Komiyama, V.N. Antonov, Appl. Phys. Lett. **93**, 073501 (2008)
23. O. Astafiev, S. Komiyama, Electron Transport in Quantum Dots, ed. by J. P. Bird (Kluwer Academic Publishers, 2003) p. 363
24. K. Ikushima, H. Sakuma, S. Komiyama, Rev. Sci. Instrum. **74**, 4209 (2003)
25. S. Komiyama, H. Sakuma, K. Ikushima, K. Hirakawa, Phys. Rev. B **73**, 045333 (2006)
26. K. Ikushima, S. Komiyama, C. R. Phys. **11**, 444 (2010)
27. W. Zawadzki, C. Chaubet, D. Dur, W. Knap, A. Raymond, Semicond. Sci. Technol. **9**, 320 (1994)
28. Y. Kawano, Y. Hisanaga, S. Komiyama, Phys. Rev. B **59**, 12537 (1999)
29. K. Ikushima, S. Komiyama, T. Ueda, K. Hirakawa, Phys. E **40**, 1026 (2008)
30. M. Henny, S. Oberholzer, C. Strunk, T. Heinzel, K. Ensslin, M. Holland, C. Schönenberger, Science **284**, 296 (1999)
31. Y. Ji, Y. Chung, D. Sprinzak, M. Heiblum, D. Mahalu, H. Shtrikman, Nature **422**, 415 (2003)
32. K. Ikushima, D. Asaoka, S. Komiyama, T. Ueda, K. Hirakawa, Phys. E **42**, 1034 (2010)
33. K. Ikushima, H. Sakuma, S. Komiyama, K. Hirakawa, Phys. Rev. Lett. **93**, 146804 (2004)
34. K. Ikushima, H. Sakuma, S. Komiyama, K. Hirakawa, Phys. Rev. B **76**, 165323 (2007)
35. K. Ikushima, S. Komiyama, T. Ueda, K. Hirakawa, Phys. E **34**, 22 (2006)
36. Y. De Wilde, F. Formanek, R. Carminati, B. Gralak, P. Lemoine, K. Joulain, J. Mulet, Y. Chen, J. Greffet, Nature **444**, 740 (2006)

Chapter 12
Material Evaluation with Optical Measurement Systems: Focusing on Terahertz Spectroscopy

Keiko Kitagishi

12.1 Introduction

The word "light" evokes the color which can be recognized with eyes, i.e., visible light in the wavelength range between 350 and 700 nm. The shorter and longer wavelength's regions are called as ultraviolet and ultrared, respectively. The wavelength of light represents the energy state, that is to say, the light at the shorter wavelength corresponds to higher energy state and the longer does to lower. Generally, the absorbance in the UV-V is region is caused by the electron transition. The absorbance in infrared region is derived from molecular vibration and rotation. Terahertz (THz) is defined as the wave of 10^{12} Hz between infrared and microwave. A lot of information about molecular structure and molecular motion has been expected from the spectra in this region [1] originating from weak energy states, i.e., lattice vibration and rotation. We, Otsuka electronics Co. LTD, developed, manufactured, and distributed various spectrophotometers including THz spectroscopic systems. I introduce, at first, our THz instruments and then other kinds of spectrophotometers.

12.2 Terahertz Spectroscopic System

12.2.1 Terahertz-Time Domain Spectroscopic System for the Transmission Measurements

We developed a THz spectroscopic system which is based upon THz-time domain spectroscopy (THz-TDS). In the THz-TDS system, a transient change of the electric field made by femtosecond laser pulse is used as a wave source and the change of

K. Kitagishi (✉)
Otsuka Electronics Co. Ltd., 3-26-3 Shodai-tajika, Hitrakata, Osaka 573-1132, Japan
e-mail: kitagishi@photal.co.jp

K. Shudo et al. (eds.), *Frontiers in Optical Methods*,
Springer Series in Optical Sciences 180, DOI: 10.1007/978-3-642-40594-5_12,
© Springer-Verlag Berlin Heidelberg 2014

Fig. 12.1 Terahertz spectrometer, TR-1000. The image is taken from [4] under permission of Vacuum Society of Japan (VSJ)

the THz pulse shape after passing through the sample is detected. The time profile (waveform) can be followed with a probe pulse by changing detection timing which the position of a sliding delay stage defines [2, 3]. Figure 12.1 shows a picture of the system we developed whose model name is TR-1000. Right side is a main body, including a femtosecond pulsed laser, laser path, THz path, a sample stage and main electronics. Left side is a control unit including power units and drivers.

The block diagram of the optic configuration was shown in Fig. 12.2. We used an Er-doped femtosecond pulsed laser of IMRA, Femtolite CS-20. It emits the laser beam of 780 and 1560 nm, simultaneously. Two kinds of THz emitters were used for this system. One is a low-temperature (LT)-grown GaAs photoconductive (PC) switch [5] and another is a nonlinear optical crystal, 4-dimethylamino-N-methyl-4- stilbazolium tosylate (DAST) [3, 6, 7]. The laser pulses of 780 nm activate LT-GaAs PC switch (dipole-type) with bias voltage of 10 kHz for the phase modulation. The laser pulses of 1560 nm activate a DAST crystal. For the phase modulation for the DAST crystal, a mechanical chopper operated at 10 kHz is used. The detection is common, an LT-GaAs PC switch (dipole-type). A half of the power of laser pulses at 780 nm is used for a probe beam to trigger the signal acquisition of THz waveform. A delay stage to shift the data acquisition point of the waveform is operated 1 mm/s for a distance of 10 mm. The stage movement is controlled with signals from the encoder with a resolution of 0.01 μm. The features of the instrument are as follows;

- Based on time-domain spectroscopy (TDS).
- Broad spectrum with two THz sources, an LT-GaAs PC switch and a DAST crystal.
- Small and desktop type. The size of the main body is 400 (W) × 540 (D) × 390 (H) mm and the weight is approximately 40 kg.
- Optical delay stage with a high speed and high resolution.
- Horizontal sample stage to enable quick measurements for various samples.
- Simple software for easy operation.

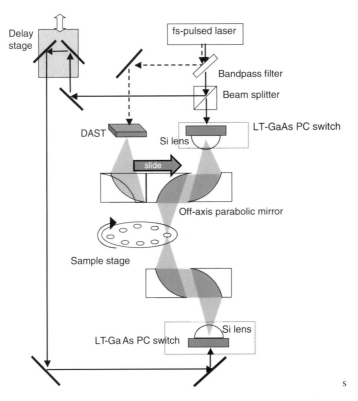

Fig. 12.2 Block diagram of optics of TR-1000. *Solid* and *dotted arrows* represents laser light path of 780 and 1560 nm, respectively

We have two types of such a kind of system configuration. The first one is that the exchange of two emitters can be manually achieved, while the second type can exchange the two automatically.

Left panels of Fig. 12.3 represent THz-waveforms from a PC switch and a DAST crystal. The acquisition time for this waveform, 10 accumulations, is approximately 2 mins. Amplitude spectra (right panels of Fig. 12.3) are obtained by Fourier transform of the waveforms. The PC switch emits the THz-wave in the lower frequency below 4 THz, while from the DAST crystal, THz-wave in the higher frequency region is radiated. By using two kinds of emitters, we can obtain the wide frequency range of THz spectra between 0.04 and 7 THz. As already mentioned, our system has a horizontal sample stage (Fig. 12.4), so liquid and **powder samples** can be measured easily as well as tablets, films, and plates. Cells for powder made by stainless have been designed for the measurements with a little amount of powder, typically 1–5 mg. In the case of powder samples, just put the sample in the cell, press lightly with the rod, and measure.

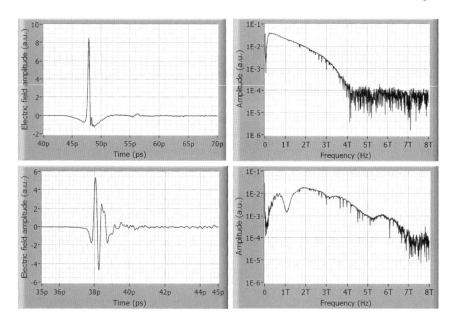

Fig. 12.3 Waveforms (*left*) and their FFT amplitude spectra (*right*) with a low-temperature-grown GaAs photoconductive switch (*upper*) and a DAST crystal (*lower*). Sliding speed of a delay stage: 1 mm/s. Frequency of phase modulation: 10 kHz. Data acquisition: 500 Hz, 5000 ch, Accumulation: 10 times. Relative humidity: approximately 5 %. The image is taken from [4] under permission of VSJ

Cells for liquid samples made by quartz have been developed. Sample liquid is put in the small cavity at the center of the cell and covered with a cover glass. The cell is put in the cell holder.

An example of the evaluation by THz spectroscopy for industrial materials is introduced. Perylene and its derivatives have been known as materials for an emitting layer of organic light-emitting diode (OLED). They show different characteristic absorption peaks in THz region each other (Fig. 12.5). The difference has been possibly derived from the different vibration energies in large region of the molecules and/or different intermolecular interactions of molecular geometry of their crystals. When powder cells are used for the measurements, the samples can be measured as powder without making tablets and can be recovered.

The absorption spectrum of water vapor in THz region shows many sharp peaks which resemble noise peaks. Accordingly, water vapor is often removed in the sample chamber of the THz-TDS systems by replacing the ambient air with nitrogen gas or by reducing the pressure. The waveforms and amplitude spectra in Fig. 12.3 still show small peaks due to a small amount of water vapor at the low relative humidity of 5 % by replacing the air with nitrogen gas. This advantage of low humidity in THz region indicates that THz spectroscopy is adequate for in-process measurements under vacuum condition.

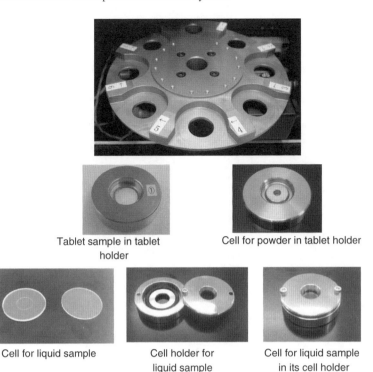

Tablet sample in tablet
holder

Cell for powder in tablet holder

Cell for liquid sample

Cell holder for
liquid sample

Cell for liquid sample
in its cell holder

Fig. 12.4 Sample stage, a variety of sample cells and cell holders

Fig. 12.5 THz absorption spectra of perylene (*black line*), perylene-3,4,9,10-tetracarboxylic dianhydride (*gray line*), N,N'-bis(2,6-dimethyl-phenyl)-perylene-3,4,9,10-tetracarboxylic diimide (*black double line*). THz source: LT-GaAs PC switch, sample amount for measurements: 4 mg each

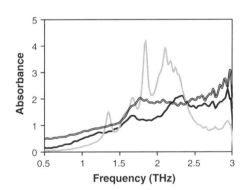

12.2.2 Terahertz-Time Domain Spectroscopic System for the Reflection Measurements

Recently, we developed the third type of the instrument. The third one includes the third emitter for the reflection measurements (Fig. 12.6). For the reflection measurements, the laser pulses of 780 nm are brought to the third emitter, an LT-GaAs PC

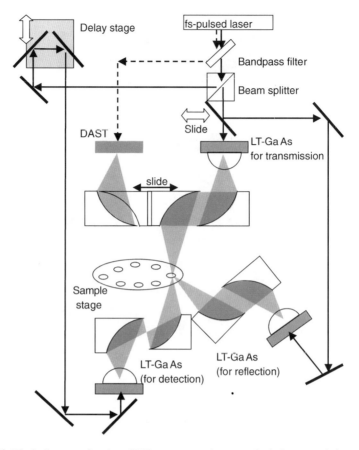

Fig. 12.6 Block diagram of optics of THz spectroscopic system both for transmission measurements and reflection ones. *Solid* and *dotted arrows* represents laser light path of 780 and 1560 nm, respectively

switch, by the insertion of a mirror and the detector is common. The sample stage for the reflection measurements leans a little. The angle of the emission and the detection is 15 degrees from the perpendicular plane of the sample.

Figure 12.7 shows an example of the transmission and reflection measurements with the system. The sample is a wire grid commercially available (Murata Manufacturing Co. Ltd. MWG40) and its specification on the catalog is 0.01–1 THz. The wire grid was set as parallel to an electric field (a hindering position) and as perpendicular (a transmission position). Transmission spectra show that THz-wave below 2 THz passes the wire grid at the transmission position, while it interrupts at the hindering position. Reflection spectra, we have almost no reflection wave at the transmission position, but 60–70 % reflection at the hindering position below 1.5 THz. It means that the wire grid measured is effective as a polarizing filter in the range below 1.5 THz.

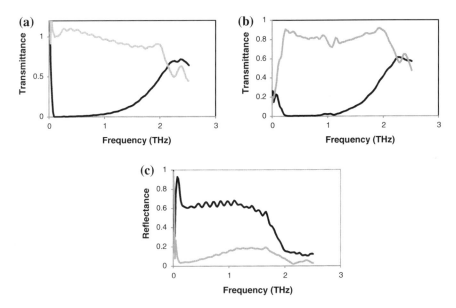

Fig. 12.7 THz spectra of a wire grid at the hindering direction (*black lines*) and transmission direction (*gray lines*). **a** Transmission spectra by using an LT-GaAs, **b** transmission spectra by using a DAST crystal, **c** reflection spectra by using an LT-GaAs

12.2.3 Optical Modification for the Transmission Measurements of the Samples with High Refractive Indices

When we measure the materials having high refractive indices in THz region such as high-resistance silicon plates with a THz spectroscopic system, the configuration of optics should be carefully constructed. The data of waveforms and transmission spectra of high-resistance Si plates are shown in Fig. 12.8, measured with the system configuration such as Fig. 12.2. The signal intensity of the waveform decreases as the thickness of the Si plate increases, though it should be constant due to the transparency of the plates in THz region. The transmission spectra are different between the three despite of the same primary material.

We assumed that the strange data of thick Si plates arose from the optical configuration in which the THz flux was focused at the sample. The ray traces before and after the sample might be symmetric when the thickness of the sample is neglected (left panel of Fig. 12.9). On the other hand, the traces are asymmetric when the sample is thick as the right panel of Fig. 12.9. Consequently, THz flux cannot be focused on the gap of dipole PC switch of the detector as the Si plate becomes thicker and the signal intensity of the waveform may seriously decrease.

It is thought that the properties were estimated adequately when THz waves radiate in parallel flux at the sample, while those were improper with focused waves. In order to evaluate the optical constants of the materials having high refractive indices in

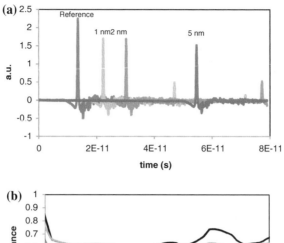

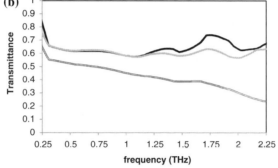

Fig. 12.8 Waveforms (**a**) and transmission spectra (**b**) of high-resistance Si plates of the thicknesses of 1, 2, and 5 mm at the focused configuration. Waveforms : Thicknesses of Si plates are shown in the plots in (**a**). *Black, gray*, and *double lines* represent 1, 2, and 5 mm thick plates, respectively, in (**b**)

Fig. 12.9 The ray traces before and after the sample. **a** thin sample; **b** thick sample. The *gray square* represents the sample having high refractive index. The *black arrows* symbolize the THz rays in dry air and in the sample

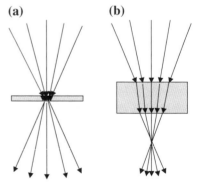

THz region exactly, a THz-TDS instrumental set-up has been modified. The THz-TDS system was designed to easily exchange the parallel/focused configuration. The second and the third off-axis parabolic mirrors are exchanged to prism mirrors

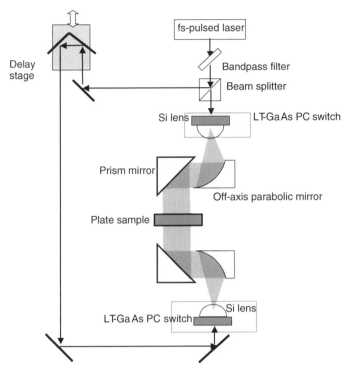

Fig. 12.10 Block diagram of optics of THz spectroscopic system for the parallel configuration for *thick samples* having high refractive indices

(Fig. 12.10). In the parallel configuration, the peak intensity of the waveform is almost the same among the three plates and the transmission spectra show no frequency dependence (Fig. 12.11).

Table 12.1 summarizes the results for Si plates having high refractive indices. In focused configuration, the thick-dependence of complex refractive indices was recognized. In parallel configuration, little thickness dependence was recognized on complex refractive index, even for a 5 mm-plate, as n and k are constant.

12.3 Material Evaluation with Visible, NIR, and Light-Scattering Spectrophotometers

The light in the region of UV, Vis, and NIR are widely used for material evaluation. Typical examples are briefly mentioned in this section with such kinds of instruments commercially available which we have developed.

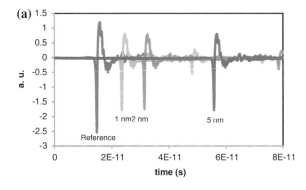

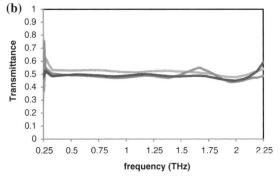

Fig. 12.11 Waveforms (**a**) and transmission spectra (**b**) of high-resistance Si plates of the thicknesses of 1, 2, and 5 mm at the parallel configuration. Thicknesses of Si plates are shown in the plot (**a**). *Black, gray, and double lines* represent 1, 2, and 5 mm thick plates, respectively, in (**b**)

Table 12.1 Complex refractive indices (refractive index n; extinction coefficient k) of Si plates calculated from the THz waveforms

Thickness (mm)	Focused configuration		Parallel configuration	
	n	k	n	k
1	3.58	<0.001	3.56	<0.001
2	3.50	<0.001	3.49	<0.001
5	2.55	0.003	3.51	<0.001

12.3.1 Color Evaluation and Estimation of Membrane Thickness with a Fiber-Coupled Spectrophotometer in the Visible and NIR Region

Fiber-coupled spectrophotometers are suitable for process monitoring at the production site. They are used for color evaluations of various display panels, i.e., plasma displays, liquid crystal displays, etc., and light sources, i.e., LED and OLED. Other industrial applications are assessments of micro-optical elements such as dichroic mirrors, microlenses, and prisms. The spectrophotometers can be used for the membrane thickness monitors and optical constant analyses of multilayer films as optical interfero-type film-thickness monitors based on reflected ultraviolet to near-infrared light. The ellipsometry can be applied for the estimation of thickness of ultrathin films and multilayers down to 0.1 nm by using ellipso-parameters.

12.3.2 Inspection of Gas Regarding Production of Semiconductors with FT-IR

The Fourier-transform infrared (FT-IR) spectrometer we developed is specialized for the gas inspection, using a Michelson interferometer and a DLaTGS detector. It can estimate the efficiency of the decomposition and the displacement volume of polyfluorocarbon gas discharged during the production process. Most processes of the production of semiconductors are achieved under vacuum, in order to avoid the formation of oxidized membrane and the adhesion of impurities. The FT-IR instrument is useful to analyze the contents of moisture and trace impurities in corrosive gas.

12.3.3 Light-Scattering Spectrophotometers

Dynamic light scattering can be used to determine the size distribution profile of small particles in suspension or polymers in solution. We have fiber-coupled dynamic light scattering systems to estimate the particle size between 1 nm and 5 μm in the wide concentration range of 0.001–10 % (w/w). It is useful to monitor the size and dispersion state of the polishing particles. The combination of electrophoresis and monitoring with light scattering method provides the information of electrical charge of the nanoparticles. Such a kind of instrument is called as zeta-potential analyzer. The zeta-potential of the surface of sheets or plates can be also measured, when the standard particle solutions are used to estimate the electro-osmotic flow. This system is used for the application researches for fuel cells (carbon nanotube, fullerene, nano metel, etc.). It can be applied to the semiconductor field, regarding researches for clarification of mechanism for deposition of foreign matters on silicon wafer surfaces, and interactions among abrasives, additives, and wafer surface.

References

1. N. Nagai, R. Kumazawa, R. Fukasawa, Chem. Phys. Lett. **413**, 495–500 (2005)
2. K. Sakai (ed.), Terahertz Optoelectronics (Springer-Verlag GmbH, Berlin, 2005)
3. M. Tonouchi, Nat. Photonics. **1**, 97–105 (2007)

4. J. Vac, Otsuka Electronics Co. Ltd. (written by K. Kitagishi). Soc. Jpn. **53**, 344–346 (2010)
5. M. Tani, S. Matsuura, K. Sakai, S. Nakashima, Appl. Opt. **36**, 7853–7859 (1997)
6. M. Yoshimura, M. Takagi, Y. Takahashi, S. Onduka, S. Brahadeeswaran, Y. Mori, T. Sasaki, M. Suzuki, M. Tonouchi. International workshop on terahertz technology, 17PS–24 (2005)
7. A. Schneider, M. Neis, M. Stillhart, B. Ruiz, R.U.A. Khan, P. Günter, J. Opt. Soc. Am. B **23**, 1822–1835 (2006)

Index

K. Shudo et al. (eds.), *Frontiers in Optical Methods*,
Springer Series in Optical Sciences 180, DOI: 10.1007/978-3-642-40594-5,
© Springer-Verlag Berlin Heidelberg 2014

Printed by Printforce, the Netherlands